design

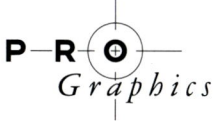

This book is to be returned on or before
the last date stamped below.

LIBREX —

RINGWOOD SCHOOL LIBRARY

RotoVision

Packaging Design

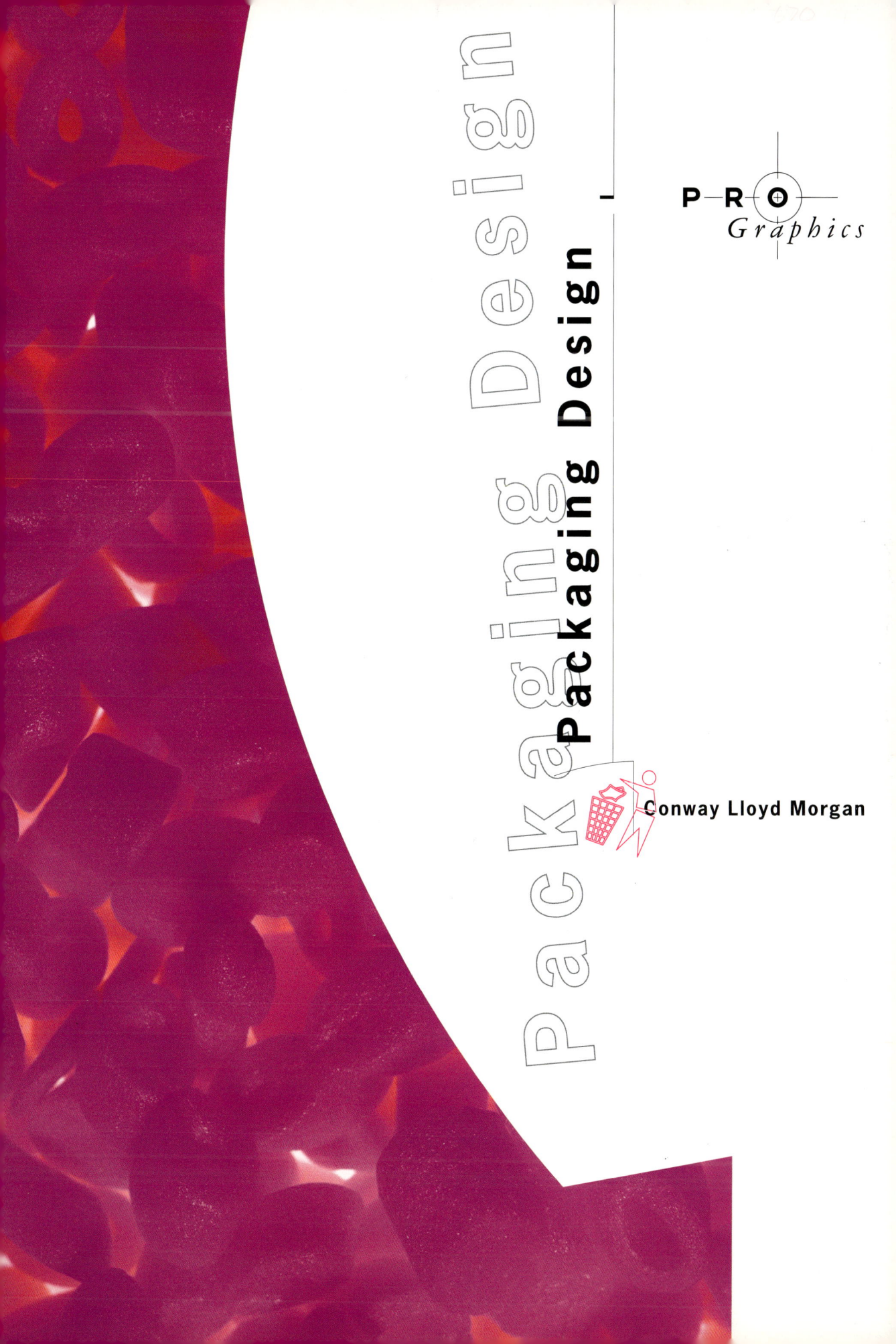

PRO Graphics

Conway Lloyd Morgan

A RotoVision Book
Published and Distributed by RotoVision SA
Rue Du Bugnon 7
1299 Crans-Pres-Celigny
Switzerland

Tel: + 41 (22) 776 0511
Fax: + 41 (22) 776 0889

RotoVision SA, Sales & Production Office
Sheridan House, 112/116A Western Road
HOVE BN3 1DD, England

Tel: +44 (0) 1273 72 72 68
Fax: +44 (0) 1273 72 72 69

Distributed to the trade in the United States:
Watson-Guptill Publications
1515 Broadway
New York, NY 10036

Copyright © RotoVision SA 1997

All rights reserved. No part of this publication may be reproduced, stored in a retrieval system or transmitted in any form or by any means, electronic, mechanical, photocopying, recording or otherwise, without permission of the copyright holder.

The products, trademarks, logos and proprietary shapes used in this book are copyrighted or otherwise protected by legislation and cannot be reproduced without the permission of the holder of the rights. The products reproduced in these pages are not necessarily available in all markets and product specifications may be changed by the manufacturer without notice.

ISBN 2-88046-262-2

Book design by The Design Revolution

Production and separations in Singapore by ProVision Pte. Ltd.
Tel: +65 334 7720
Fax: +56 334 7721

Packaging  Design

Acknowledgments

The creation of this book was made possible – and pleasurable – by the enthusiasm and co-operation of many people. Patrick Farrell and Jonathan Kirk at Raymond Loewy International, Andy Quady and Cheryl Lizak at Quady's Winery, Roxanne Hale at Guiltless Gourmet, Jacob Leinenkugel at Leinenkugel's Brewery, Mark Kinger at John Brady Design, Ian Welsh at Safeway Stores, Geoff Giles at SmithKline Beecham, Davide Nicosia and Maurits Pesch at Nicosia Creative Expresso, Mike Green at Pecos Design, Hans Kan at Kan Creative Consultants, Jimmy Yang and Max du Bois at The Identica Partnership, Gert Boven at Lumiance, John Miescher at SLI, Graham Scott for Adelphi Whisky, Andrea Yanowitz at Fractal Design, Jackie McQuillan at Virgin Cola, Jo Moore at Lewis Moberly, Hans Riechmann at Keggy, Harry Pearce at Lippa Pearce, David Foster at Design Partners, Jon Turner at the Body Shop, Chris Jones at Tomy UK, Peter Foskett and Gavin Thomson at Pentagram, Lee Ellis (for the flat plans) and Natalia Price-Cabrera alll helped, and I'm very grateful to them.

Special thanks go to Julian Busselle for his splendid photography, Mark Roberts at The Design Revolution for the creative visual interpretation, and Angie Patchell for vigorously shepherding the book through its final stages.

Finally, a tribute to Jean Koefoed. Not only was he instrumental in getting this book published, he has also offered me a wealth of willing advice and active encouragement during the last fifteen years, whether over a memorable mint julep in Maryland, or in the raucous aisles of the Frankfurt Book Fair. One of the first to understand deeply and intuitively the true qualities of design publishing, Jean shared his beliefs and experience with enthusiasm and wit. This book is dedicated to his memory.

Conway Lloyd Morgan, London, January 1997

Contents

ACKNOWLEDGMENTS — 8

INTRODUCTION — 8

THE DESIGN BRIEF — 12

DESIGN STUDY: — 16

- **Safeway Stores own products** — 16
- **Safeway Savers** — *Low cost but high value* — 18
- **Safeway Cyclon** — *Brand within an own brand* — 22
- **Safeways Premium Ice-Cream** — *Looking the taste* — 24

FOOD PACKAGING — 26

- **Toblerone** — *Maintaining tradition* — 28
- **Yoplait Yoghurt** — *From health to luxury* — 30
- **Horai Dairy Products** — *Bringing the farm to the kitchen* — 34
- **Panem** — *Wrapping up a new project*
- **Guiltless Gourmet Taco Chips** — *Selling on color* — 38

DRINK PACKAGING — 42

- **Sainsbury's Vin Rouge de France** — *Material advantages* — 44
- **Lucozade Sport & Lucozade Energy** — *From bedside to sports field* — 46
- **Adelphi Whisky** — *Marketing exclusivity* — 52
- **Smirnoff Vodka** — *Premium celebration* — 56
- **Virgin Cola Pammy** — *Profiling Cola* — 60
- **Quady's Winery** — *New World Sweetners* — 62
- **Leinenkugel's Beers** — *North woods to world wide web* — 70

LEISURE & GIFTS PACKAGING		**72**
Nicole Fahri	*The shopping bag rules*	**74**
Baby Vision Toys	*Creative Play*	**76**
Sellotape	*Refixing a household name*	**82**
Orient Tea Rooms	*Crossing the tea*	**86**
Metroblade Inline Skates	*Graphics with attitude*	**88**
COSMETICS & BEAUTY PACKAGING		**94**
Benetton Tribu Gift Set	*United and international*	**96**
Body Shop Bodycare Products	*Keeping holistically ahead*	**102**
Insignia	*After shave for Robocop*	**108**
Sempre	*Putting a bow on it*	**110**
HEALTHCARE & HOUSEWARE		**114**
Toilet Duck	*Getting into shape*	**116**
Oropharma	*Racing into health*	**118**
Simoniz	*Bottling it up*	**122**
TECHNOLOGY PACKAGING		**126**
Painter Software	*Keeping a lid on it*	**128**
Sylvania Lighting	*Recycling the polar curve*	**130**
Olympic Weathering Stain	*Painting the difference*	**136**
Keggy	*Engineering innovation*	**140**
CONCLUSION		**144**
Halfords Motor Oil	*Design with bottle*	**146**
Reading List		*156*
List of Designers		*157*
Index		*158*

It has often been said that the packaging is the best advertisement for a product: in today's competitive retail environment it is even more true. Good packaging design is a key part of retail success. This book looks at examples of packaging, in different fields and with different materials, supported by explanations of how the final design was achieved. In this introduction we will look at the skills you, as a designer, need to develop to create good packaging design, and why the packaging design business is so important, and how packaging fits into the development and promotion of products.

PACKAGING DESIGN IS GRAPHIC DESIGN

The basic graphic design skills – the use of color and type, the right sense of balance and proportion, the choice of materials and finishes – all apply to packaging design. But packaging design is three-dimensional, and is applied to bags and boxes, cans and bottles, as well as to flat surfaces. So there are other, more specific skills: learning how to print on glass and metal, on curved or flexible surfaces, for example, and understanding the mechanics of plastics, paper and card so that the design will fold or cut out properly. Most packaging design is intended for medium- and long-run manufacture, and you need to consider the feasibility of producing any design economically.

Introduction

PACKAGING DESIGN IS IMMEDIATE DESIGN

As you walk through a shop or supermarket, your eyes skim over the goods on display. Each product has at most a half-second or less in which to make its claim for recognition to the customer. So the packaging design statement needs to be direct. This doesn't mean that it must be loud, or simple, but it must be clear to the audience for which it is intended.

PACKAGING DESIGN IS CLIENT-LED DESIGN

New packaging designs are commissioned for a number of reasons: the product may be wholly new, but more often the product is to be added to an existing range, or the packaging of an existing product is being reworked. Whatever the motive for the design, understanding the client's needs as fully as possible is essential. If the product is new, what is the target market? If the product is an addition, what are the strengths of the existing brand packaging which need to be maintained? If the product is being redesigned, is this for a new market, or to revive falling sales, and if so what are the strong and weak points in the present design that need to be watched? The fuller the brief from the client, the better the end design will be.

Introduction

PACKAGING DESIGN IS COMPETITIVE DESIGN

Modern products are competing for shelf space in shops and supermarkets, as well as for the attention of the customer. Successful packaging design needs to be based on thorough research, not only on the design of competitors' products, but also on how the product category is displayed and sold. This would include stacking methods, ambient lighting, storage, transport and security.

PACKAGING DESIGN IS ADVERTISING

Packaged products are not independent: they are supported by advertising and promotion (commonly, for example, new products are offered at a discount). The design must share and reflect the aims of the advertising program, for example through slogans, colors or images. The packaging must also follow the marketing plan in addressing the target audience.

PACKAGING DESIGN IS TEAMWORK

The designer is part of a team on any packaging project. The 'main client' may be the product manager, but the designer will also have to work with the marketing director, the advertising agents, and the people responsible for production and distribution. For a major project, other experts will be involved: market researchers, paper engineers, color consultants, finance directors, and so on. Working with and within the team is a major key to success.

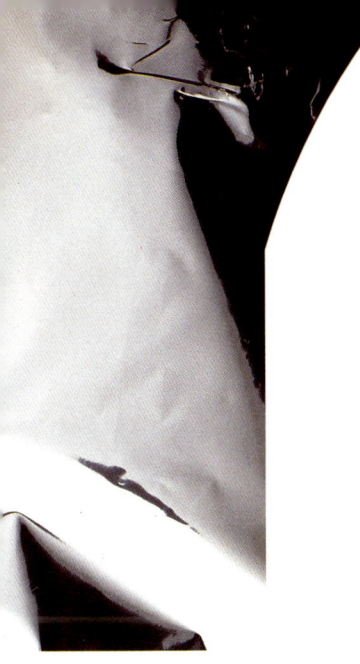

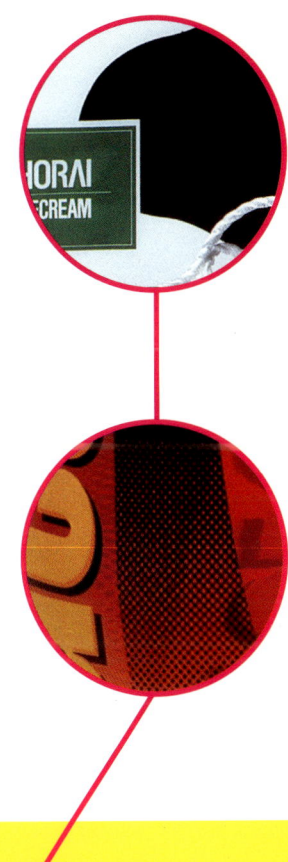

tion

duction

GETTING ON FROM HERE

On the following pages you will find a sample design brief form: it sets out points to check in developing a design from the client briefing, with an analysis of the categories involved. This is not a formula but a model: feel free to adapt it to your own needs. The same elements have been used to present the information on each of the case study designs in the rest of the book, and how this works is shown in detail on page 14. Where a case study runs over several pages, the link buttons on the first page will show you where additional information is to be found. The case studies have been selected from the best design practice around the world so that you can see how other designers have dealt with particular problems, and the creative solutions they have found.

As an introduction to the general task of packaging design, you will find on page 16 an in-depth review of part of the design production of a major supermarket chain, as an illustration of the challenges and complexities of contemporary packaging both for designers and managers.

DESIGN BRIEF

Job no | Client job no | Date

Client name | Contact
Address
Phone | Fax | e-mail

PRODUCT
Name | (Provisional/actual)

Product type

Weight | Dimensions | Material

Packaging type card/paper/box/can/bottle/sealed plastic/vacuum formed/other

Proposed packaging | Dimensions | Weight

Special sealing requirements Yes/No Bulk packing type

DESIGN TEAM in house

CLIENT TEAM
Name | Position | Contact number | Fax

SCHEDULE
Research | Preliminary | Final rough | Production date

DESIGN
New | Add to brand | Revise | Revise brand

OTHER PRODUCTS IN BRAND
Fixed elements | Logo | Brand colors | Typefaces

TYPEMATTER
Main text | Subsidiary text | Company details
Ingredients list | Sell-by date | Price | Statutory information

MARKET
Age range | Sex | Social group | County | Other

COMPETING PRODUCTS
Market research | Existing | New Market testing

ADVERTISING
Budget | Medium | Campaign length | Style
Point-of-sale | Display | Other

CLIENT KEYWORDS 1 | 2 | 3
4 | 5 | 6

Date of briefing | Date of next meeting | signed

12/13

DESIGN BRIEF NOTES

The form on the facing page lists the elements to be considered as part of a design brief. It is a way of guiding your thinking and dealing with the client. Not all these categories will be valid for every project, nor will all the information fit onto one sheet. Use this as a header, and list additional information (for example typematter, a list of competitors' products and comments or research on them, etc.) on subsequent sheets. Update the header sheet at each meeting, and keep all the relevant sheets at the top of the job file.

This is the key decision on materials. It will be affected by the manufacturing process available, by the intended method of sale (on the shelf, on hooks, by mail order, and so on), as well as by your own choices.

These are the fixed elements that reach across the brand to the whole identity of the company and its product range. Working out how they are used by the client, and so studying the corporate identity of the client, is very important. Never think of logos or brand colors as 'something that has to go on', they are often the strongest selling point a product has got, and the client has invested years in building the confidence of the market-place in the name.

The product type is often the first parameter for packaging: is it a liquid, a powder, a manufactured object? Is it soft or hard, sensitive to light or heat? And so forth. If it is a food product, or a personal care one, or if it is a leisure or sports item, you know where to start looking for the competition.

You should be aware of the various methods for sealing products, either to preserve them or to prevent tampering or leakage. The child-proof lids on pharmaceutical products, for example, change the shape of bottles and jars considerably. Security stripes on wrapped foods need to be clearly distinct from the overall design.

Designing the packaging for a wholly new product is a tough and exciting challenge. You need to work hard at researching the market and the competition, and devising a solution that will 'sell itself off the shelf'.
Adding products to a brand is just as challenging, but requires a more disciplined approach. You will have to learn to read the key features of a brand and apply them to the new product.
Revising an item needs to be led by the client: what are the reasons for the revision? (A new formula, a new title, or a new size of packaging could be among them.)
Revising a whole brand is a key exercise in maintaining a product's commercial life, the successful aspects of the old design need to be carefully considered, and the alternative strategies well researched.

Another key question: who is it for? The answer will affect the choice of colors, the kind, and size of typefaces, the illustrations used and the packaging material. By respecting these questions the final product will fit more clearly into the marketing plan and the promotion for the product.

The packaging and advertising of a product must work together. Not only in terms of point of sale displays that remind the customer about the advertising while offering the product, but the whole direction of an advertising campaign needs to be consolidated by the look of the product, in terms of the market addressed, and the semantics of the packaging. A full briefing on the advertising and marketing strategy is essential for a successful product and brand design.

It is good practice to identify key words or phrases that describe the product for the client. Often these will come from the marketing and advertising proposals: is this product serious or fun, for example, hot or cool, brash or sophisticated? This ties in with the intended market and price range, and the way in which it is to be sold.

HOW TO USE THIS BOOK

KEY INFORMATION
In this book we have adopted a version of the design brief to describe the designs. The product name, designer and client: the key partnership in any project.

BRIEF HEADINGS
The type identifies whether the design is for a new product, a redesign, an addition to a brand or a completely new brand. The time shown is the development time: from initial brief to the start of manufacturing or launch. The market is the target market for the product, while support describes the advertising and point of sale material developed to promote the product. The key words are a shorthand description of the intended placement of the product.

FOOD

TYPE	Redesign of existing range
PRODUCT	Dairy products
MATERIAL	Printed card and plastic
SIZE	Various
TIME	7 months
ELEMENTS	Barcode and company logo
MARKET	General adult
SUPPORT	In-store promotions
KEY WORDS	Fresh, countryside, direct

DESIGN BRIEF

The brief was to design a new packaging range for dairy products: milk, cheese, ice-cream, yoghurt and honey. The design should stress the fresh, natural taste of the product, and be recognizable as a product group in different sections of the supermarket. For sales through specialist shops, bags and chilled containers were also required. The design was introduced in 1992, and market research suggests it has gained strong popular identification.

The main design motif is the black and white pattern of a Jersey cow, with green (for freshness and grass) as a main subsidiary color, and other strong colors to distinguish products within each range. Products to be sold shelf-out (such as honey) are labelled from the side. Those that can be taken from a freezer chest (ice-cream) carry strong top labels.

Horai Dairy Products

DESIGNER Kojitani, Irie & Inc. Tokyo, Japan
CLIENT Horai Co, Tokyo, Japan

3 34/35

'The products have to work as a range, even if placed in different locations in the store: they also have to be recognisable on the shelf at home.'

DESIGN The Jersey cow has also been used as a design motif more recently, for computer packaging, by the successful mail order company Gateway 2000 (whose operations began on a farm). This highlights the difference between short- and long-term packaging: the computer box is thrown away or stored once the computer is unpacked, while the packaging for dairy products stays on the shelf or in the refrigerator until the product is used up. A computer finished in white and green would be fun, but would have no meaning: dairy goods easily identifiable by a countryside motif are a different matter.

Key Factors

DESIGN
The new brand had to link different products, and make a core statement 'farm fresh' in a busy shop setting.

P.34

DESIGN BRIEF
The design brief is analyzed in bold type, with further discussion and comments in standard type.

KEY FACTORS
The buttons identify special aspects of the design and guide you onto further discussion of the case study on the pages shown.

FOOD

TYPE OF DESIGN

By 1994 a change was needed marketing director Jonathan K Berry, decided to retain certain and the color illustrations of fr these in a more luxurious way cardboard housing for multipac underscored on the twinpacks labelling so that the name appe the illustration once on each si than making the product image repeating the same design acr The new packaging has increas sales volume by ten per cent i highly competitive market.

The earlier packaging (below) separated name, description and illustration into bands. The new pack links all three visually.

Yoplait Yoghurt

2 32/33

The brief was to reinf
appetite, app
values t

FULL COLOR IMAGES

The main image shows the product in its complete design livery.

CAPTIONS

Specific aspects of the design solution are listed in the captions.

With the change in marketing to multipacks, the labelling system was altered so as to show both text and image clearly, side on.

CONTINUATION TEXT

The issues raised by the design are elaborated upon here, using the key factor headings.

ELEMENTS

A four-pack of yoghurt normally has the 'outside' semicircle of the label facing out of the pack, but Roger Berry realized that this was irrelevant once the yoghurts were separated. Better to reinforce the pleasure of the product by alternating the label positioning. So on the shelf, the consumer sees both the text with the product name and the illustration, rather than two text labels. A very simple change, involving no cost, doubles the message sent by the product to the purchaser.

DEVELOPMENT

On subsequent pages, key factors, including the extension of the design across a range, are discussed.

The move from a formal 'healthy' typestyle to a more exotic one reinforces the market shift of the product.

e products and **premium** ntain **market** leadership.

QUOTES

These are remarks by the designer or client on the purpose or success of the design, or comments by other experts or writers relevant to the design.

ADDITIONAL IMAGES

These show specific aspects of the design, or related products or design stages.

Safeway

Design Study

Packaging design is a key factor in many industries, but within the food and drink industries it is certainly one of the most important. With the increasing move worldwide to supermarket shopping, how supermarkets handle their in-house design provides an excellent starting point. The Safeway chain in the UK began as a franchise of the American store, but is now independently owned. They have over 300 stores and superstores offering a complete range of food, drink, groceries and housewares. Their own brands represent a considerable portion of their turnover, sold in competition to other manufacturers' brands.

According to Ian Welsh, packaging design manager at Safeway's head office in the London suburbs, own-brand products are pitched at three levels: basic products to meet everyday household needs; standard products that cover a wider range and premium products aimed at a more indulgent, discretionary market. Within these broad categories, individual product groups can be planned as brands within the brand. Each such brand or design, Welsh says, has to maintain its individual presence in the market, and support the overall Safeway image. On the

own Brands

following pages we look at three groups of Safeway own products: Safeway Savers, a basic low-cost, good-value line; Cyclon washing powder and liquid, an own-brand everyday product; and a premium range of luxury ice-cream.

Safeway source all their design work externally, employing a team of in-house design executives who supervise the development, introduction and management of different product areas. The design management function needs to combine design awareness with management skills: some, but not all, of the executives have design experience, and their main work involves liaising between the central strategic unit and the managers developing individual products, with briefs and concepts being approved by both sides. Getting supplies of new packaging produced to coincide with the sales cycle is a main issue, according to Welsh. Often products come from a number of suppliers, and their production has to be coordinated and the quality controlled. That is the technical aspect: positioning the products correctly is also a major issue. 'We are always looking for classic design which will last and be built on in the future, compared to a situation some years ago when more transitory design was acceptable.'

Safeway Savers

Designer Wagstaffs, London, UK
Client Safeway Stores plc, Hayes, UK

Type	New product range
Product	Basic foodstuffs and housewares
Material	Various
Size	Various
Time	14 months
Elements	Barcode and company logo, endorsement
Market	General
Support	In store promotion
Key words	Value, quality,

Design Study

Safeway Savers is a range of basic food and home products including fresh vegetables, tinned goods, frozen vegetables and meat, breakfast foods, bread, juice and soft drinks, milk, and household cleaners, among others. This range was first marketed in 1994, and has proved a considerable success in offering low-cost, good quality products to the public. The products are packed in a variety of materials, but use uniformly white lettering on a red and green background. These are the company's house colors. The word Safeway follows the house style, with a wave shape undercutting the lettering, while 'savers' is in a bold swash italic face.

'*If you saw a lemonade bottle on a beach at a distance, you'd know it was ours: that was the aim of the design.*'

The Savers logo incorporates both the main Safeway logo, with its undercurve, and the word Savers, reversed out of red.

Key Factors

Market
The Savers product range emphsizes value not low cost: this must show through in the design.

P.20

Elements
The design had to use Safeways house colors and be seen across the whole store in different locations.

P.20

Materials
The design solution had to be applicable to a wide range of materials: paper, plastic and card, among others.

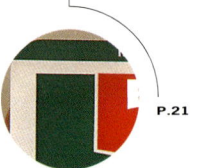
P.21

This seems a simple design solution, but is actually complex both in planning and execution. 'The brief was fairly simple: it had to reflect Safeway; it had to be seen as a basic value option; it had to cross endless packs and processes without difficulty; it had to be visible. But within that it had to be positioned very accurately: it mustn't look too cheap. In fact, it has to show we were proud of it, because it was individually ours,' according to Ian Welsh.

TYPE OF DESIGN

'These kind of projects are extremely tough. You have no margins of flexibility, no visual imagery to play with whatsoever. It has to go from a small to a very large label, for example, and be produced by a wide range of different printers and manufacturers.' These constraints make the job of planning such a range an exacting one, down to the title chosen: 'savers' suggests judgement and value, for example, while 'economy' or 'bargain' might suggest penny-pinching or poor quality.

Bread, tea, juice, yoghurt, tins: the design message needs to be consistent across them all, despite size and material changes.

SAFEWAY Savers

Designer	Wagstaffs, London, UK
Client	Safeway Stores plc, Hayes, UK
Elements	Barcode, logo, ingredients list, guarantee, product promise
Material	Various

Design Development

The Savers collection shows the range of packaging materials commonly met in food packaging: glass and PET plastic bottles for liquids, tin cans for foods, transparent polythene bags for fresh produce, sealed polythene bags for frozen foods, cardboard boxes for dry goods and plastic containers for yoghurts and liquids. To emphasize that this is a value range rather than an inexpensive one, the same packaging materials are employed as for standard products: Tetra packs are used for milk and orange juice and the cooked ham is in a modern, resealable polythene pack.

The pack must invite the consumer to purchase: too simple a design would fail, as would one too elaborate.

If you saw a **lemonade bottle** on a beach at a distance, you'd know it was ours: that was the aim of **the design**

Materials

The production processes for labelling are also standardized: two color printing (red and green) on paper labels for bottles and tins or screenprinting on clear polythene in three colors (red, white and green) or in two colors on white plastics. More sophisticated techniques (screenprinting onto glass, embossing or blind-stamping) have been rejected because they would be inappropriate to the central marketing proposal of the range, that the value is inside the packaging.

Safeway Cyclon

Designer Wagstaffs, London, UK
Client Safeway, Hayes, UK

'Creating a sub-brand requires more resources and attention than adding a new product to the own-brand range. But it is a powerful endorsement of a new product.'

Type	Creation of new brand within a brand
Product	Washing liquid and powder
Material	Plastic bottles and bags, card boxes
Size	Various
Time	24 months
Elements	Barcode, main logo, brand logo, guarantees
Market	General
Support	In-store promotion
Key words	Powerful, new, efficient

Design Study

A new type of product always creates a challenge for those developing house brands. There is a technical problem of sourcing a suitable product (though Welsh points out that a number of manufacturers are now specializing in producing 'own-brand' products only for supermarket chains, and in developing new products for this market). The design problem posed is to produce a design that is both independent and part of the wider corporate brand, and carries the same message and values as the target market, without appearing to be a look-alike or copy. Sometimes the way to achieve this is to launch a new product as a 'brand within a brand'. This was the choice made for Safeway Cyclon, a range of washing liquids and powders that incorporated new chemical technologies suitable for modern washing machines.

The advantages of a brand-within-a-brand are that the product has its own name and so its own identity therefore making it is easier to promote. But the drawback is that it also requires constant attention and updating, and that it may deflect attention from the main brand. Therefore, Welsh points out, sub-brands like Cyclon should always be used strategically, with the necessary resources behind them. Even though the supermarket has the advantage of controlling its own shelf-space, this cannot be overexploited by too many sub-brands, simply because they are not the resources to manage each fully.

Key Factors

Type

A brand-within-a-brand requires careful placement, and proper resources for development.

P.22

Materials

A new brand needs to look to creating its own equities, for example through a bottle or pack shape, as here.

P.23

The logo for a brand-within-a-brand needs to work alongside the main brand.

Packaging design for washing powder has moved from plain whites and natural colors in the early 1980s to a much more sophisticated color range. Within these, Cyclon has both to compete on equal terms and make an individual statement. The process of designing the actual graphics began with the decision to launch the product, and has gone through a number of revisions, and will continue to be revised to maintain the product's place in the market. To date, however, the product has been successfully launched, and is holding its place well.

The product range covers boxed loose powder, bottled liquid and liquid in refill pouches.
The bottle shape is an individual design, and so can be listed among the design equities along with the name and color schemes used to differentiate varied applications such as color, biological and non-biological washes.

Safeway Premium Ice-Cream

Designer Wagstaffs, London, UK
Client Safeway Stores plc, Hayes, UK

'We are always seeking for classic design, that will last.'

Product	Luxury ice creams
Material	Plastic tubs with printed paper labels
Size	250 ml tubs
Time	4 months
Type	New product line within own-brand
Elements	Barcode, ingredients, guarantees and company logo
Market	General
Support	In-store promotions
Key words	New, luxury, original

Design Study

Type positioning and placement needs to meet in store use, but the choice of typeface depends on the product type.

Ice-cream has moved from being party food for children, or a summer treat, into serious adult indulgence food. The success of brands such as Häagen Dazs and Ben and Jerrys in creating new flavors and a new image for the product is one evidence of this, another is the development of individual ice cream products based on confectionery, such as Mars and Snickers.

When Safeway were looking to get into this market with an own-label premium ice-cream, they realized that it had to meet the consumer's expectations in two ways: firstly by the choice of flavors, and secondly through the packaging. Square white plastic boxes with labels were abandoned in favor of small round tubs, in this case in matt black plastic, an echo of the round waxed paper tubs used by the main brands. As the ice-cream in the store freezer is normally seen from above, the design of the main top label was the most important factor. The main rival brands have strong visual identities (Häagen Dazs use gold and brown typography on white, Ben and Jerry's a trademark drawing). Safeway opted for strong colors (reflecting the flavors) and lush photography of the main fruit ingredient. These indulgent photographs were balanced by equally evocative names such as Strawberry Scandal, Peach Passion and Vanilla Vanilla (so good we named it twice...).

Key Factors

Type

A range intended to compete with market leaders must respect but not imitate the expectations they arouse.

Design

The choice of photographic images confirms the 'luxury' product statement.

P.24

P.24

Black plastic tubs mark distance from the competition, while suggesting the daring nature of the product.

The Safeway name is prominent without being overbearing. Compared to the rapidly changing designs of a few years ago, Safeway now look for classic designs that will last. This product, which offers an original approach to a newly-established product type, is a good example of success in this approach.

1 Food

LOOKING FOR TASTE

Food packaging is one of the most important areas of packaging design for two rather basic reasons. Firstly, most foodstuffs cannot be easily sold unpackaged, certainly in the developed world, and secondly food represents an important percentage of most families' budgets each week. It is a market-place that is both essential and competitive. Food packaging has evolved from the wooden barrel and the paper bag to the almost overwhelming array of technologies for processing, packing and labeling food today, whether tinned or canned foods, frozen foods, freeze-dried foods and chilled foods.

Food packaging reflects, in a mature marketing environment, a whole range of wider social aspirations. Food as a gift (with a Toblerone chocolate package), food as an indulgence (Yoplait yoghurt), food as sustenance (Horai dairy products), food as a convenience (Panem bread), food as health plus indulgence (Guiltless Gourmet taco chips) are some of those dealt with in this section. At the same time food packaging offers an overview of the different packaging materials – glass, card, plastic, cellophane, paper and foil – which are represented here. Food packaging invites the designer to balance a whole range of factors.

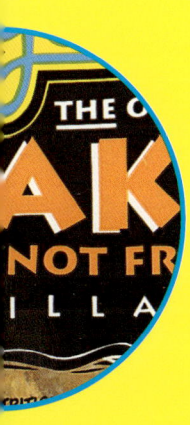

Firstly there are the problems of making food portable: the technical considerations of manufacturing and processing food and moving it around. Then it must be kept clean and safe, an area often bounded by national and international regulations. Thirdly, its nature – in the form of ingredients, cooking or preparation instructions, and dietary values – must be communicated. Finally, and of overriding importance, it must communicate with the customer's desires. It must say 'Enjoy!'

In understanding this fourth, and final area, the designer must look at three things: the history of the product itself and the manufacturer, the competition in the market-place, and the attitudes and aspirations of the client market. This process of research must begin as soon as the brief is discussed, in parallel with the technical aspects, while the motivated designer will also be independently researching the general market all the time.

FOOD

Toblerone Multiple Gift Pack

DESIGNER	Blueberry Design, London, UK
CLIENT	Kraft Jakob Suchard, Geneva, Switzerland

1 28/29

TYPE	New design
PRODUCT	Chocolate bars
MATERIAL	Printed and folded card
SIZE	6 x 50gm packs
TIME	1 month
ELEMENTS	Barcode and company logo
MARKET	General in Europe and Saudi Arabia
SUPPORT	Display bins
KEY WORDS	Special, exclusive, traditional

Design Brief

Toblerone is a milk chocolate bar studded with almonds and nougat. A Swiss product, it uses triangular segments of chocolate both to distinguish itself from the traditional flat chocolate bar, and as a visual echo of the mountains of Switzerland.

Key Factors

DESIGN

A new pack for an established product needs to maintain and develop the existing design equities.

P.29

The cutter guide shows that the complex shape can be cut economically from a single sheet.

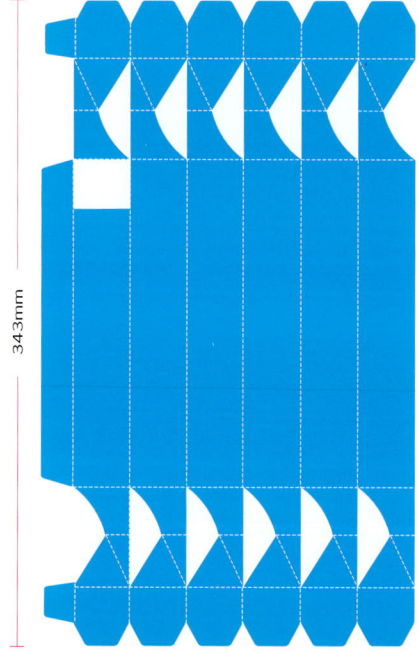

343mm

The distinctive triangular pack, printed in gold and red on a cream base, has deliberately been left untouched for decades. It is an excellent example of visual or design equity, that is to say a visual element in a product which is so synonymous with the product that it can only be changed at great risk, and also plays an important part in maintaining the product's identity with consumers. Toblerone (now owned by KJS) has deliberately chosen to maintain the triangular format, even when adding new products to the range such as Toblerone Dark.

' Our aim was to extend Toblerone's established appeal into a higher volume market.'

MATERIALS The multi-pack was developed by Blueberry Design in London. According to Will Sellers, director, the brief called for a design that would retain the visual equity of the existing single product but in a luxury format to compete on the shelf with luxury chocolates. 'We chose to emphasize the gold and red in the overall design, as well as maintaining the triangular element through a hexagonal carton.'

In order to meet changing demands other developments have been made to the packaging. These include a plastic wrapper for a smaller bar, as well as the multipacks which hold several bars in one pack. The latter was developed for the gift market, and in particular for sale in duty-free shops.

The Toblerone gift pack retains the triangular form which is a key design equity of the original product.

Yoplait Yoghurt

Designer Raymond Loewy International, London, UK
Client Waterford Foods, London, UK

Type	Redesign of existing packs
Product	Fruit yoghurt
Material	Printed plastic containers
Size	Various
Time	8 months
Elements	Barcode and company logo
Market	General
Support	TV and press advertising
Key words	Natural, indulgent, pure

Design Brief

The packaging design introduced in 1986 by Loewy, and a market leader for nine years.

A major motive force for changing packaging design, especially in the food market, is changing customer perceptions. 15 years ago yoghurts, whether plain or fruit-flavored, were perceived as simple and pure health foods. Increasing sophistication and choice in the dessert market has led to a redefinition of yoghurt as a more indulgent food. Loewy International had designed the French group Yoplait's packaging for the UK and Irish markets in 1986, and the product held its leadership position well.

'*The brief was to reinforce the product's appetite appeal and premium values, to maintain market leadership.*'

Key Factors

Design

Changing the positioning of a product involves carrying the existing values into a new context.

P.32

Elements

The redesign has to follow new sales patterns in supermarkets and shops, for multipacks rather than singles.

P.33

The new pack uses integrated photography and graphics to express the move to a more indulgent market-perception of yoghurt.

TYPE OF DESIGN

By 1994 a change was needed. Loewy, according to marketing director Jonathan Kirk and designer Roger Berry, decided to retain certain original elements (the logo and the color illustrations of fruit, for example), but present these in a more luxurious way. This was applied to the cardboard housing for multipacks, and subtly underscored on the twinpacks by reversing the labelling so that the name appears once and the illustration once on each side, rather than making the product image banal by repeating the same design across both. The new packaging has increased sales volume by ten per cent in this highly competitive market.

The earlier packaging (below) separated name, description and illustration into bands. The new pack links all three visually.

The brief was to reinforce the products **appetite, appeal** and **premium** values to maintain **market** leadership

With the change in marketing to multipacks, the labelling system was altered so as to show both text and image clearly, side on.

Elements

A four-pack of yoghurt normally has the 'outside' semicircle of the label facing out of the pack, but Roger Berry realized that this was irrelevant once the yoghurts were separated. Better to reinforce the pleasure of the product by alternating the label positioning. So on the shelf, the consumer sees both the text with the product name and the illustration, rather than two text labels. A very simple change, involving no cost, doubles the message sent by the product to the purchaser.

The move from a formal 'healthy' typestyle to a more exotic one reinforces the market shift of the product.

FOOD

Horai Dairy Products

DESIGNER Kojitani, Irie & Inc.
Tokyo, Japan
CLIENT Horai Co, Tokyo, Japan

TYPE	Redesign of existing range
PRODUCT	Dairy products
MATERIAL	Printed card and plastic
SIZE	Various
TIME	7 months
ELEMENTS	Barcode and company logo
MARKET	General adult
SUPPORT	In-store promotions
KEY WORDS	Fresh, countryside, direct

DESIGN BRIEF

The brief was to design a new packaging range for dairy products: milk, cheese, ice-cream, yoghurt and honey. The design should stress the fresh, natural taste of the product, and be recognizable as a product group in different sections of the supermarket. For sales through specialist shops, bags and chilled containers were also required. The design was introduced in 1992, and market research suggests it has gained strong popular identification.

The main design motif is the black and white pattern of a Jersey cow, with green (for freshness and grass) as a main subsidiary color, and other strong colors to distinguish products within each range. Products to be sold shelf-out (such as honey) are labelled from the side. Those that can be taken from a freezer chest (ice-cream) carry strong top labels.

' The products have to work as a range, even if placed in different locations in the store: they also have to be recognizable on the shelf at home. '

DESIGN The Jersey cow has also been used as a design motif more recently, for computer packaging, by the successful mail order company Gateway 2000 (whose operations began on a farm). This highlights the difference between short- and long-term packaging: the computer box is thrown away or stored once the computer is unpacked, while the packaging for dairy products stays on the shelf or in the refrigerator until the product is used up. A computer finished in white and green would be fun, but would have no meaning: dairy goods easily identifiable by a countryside motif are a different matter.

Key Factors

DESIGN

The new brand had to link different products, and make a core statement 'farm fresh' in a busy shop setting.

P.34

Panem

Designer Raymond Loewy International, London, UK
Client Barilla SpA, Milan, Italy

'The brand proposition is "fresh every morning". The packaging supports this.'

Type	New product range
Product	Bread
Material	Printed card and clear cellophane wrapper
Size	500gm loaf
Time	6 months
Elements	Company logo
Market	General
Support	Television and press advertising
Key words	Fresh, new

Design Brief

'Next best thing to sliced bread' is an ironical reflection on a product that has been an indicator of shopping habits for a couple of decades. The introduction of wrapped sliced bread in the USA and the UK heralded the move from individual, served shopping in small shops to the self-service of supermarkets, while today the range of breads available in this format is a guide to how much more knowledgeable consumers are about health.

In continental Europe the tradition of the local, daily baker (and other local food shops) has not been eroded to the same extent by the arrival of supermarkets, and the dietary value of bread is a fixed part of general culture. The challenge set to Patrick Farrell of Loewy International by Barilla, a major Italian food producer, was to package bread for supermarket sales.

Key Factors

Materials
The choice of a clear wrap, by maintaining visibility, helped the consumer move over to a new bread in a new setting.

P.37

Logo
This logo uses a Latin tag and a sunrise motif to emphasize continuity in a new context.

P.37

The final combination of wrap and logo was intended to echo a traditionally presented loaf of bread.

Together with his colleague Graham Spice, Farrell adopted a solution that would allow the quality of the product to show through. The outer wrapping was totally transparent polythene, fairly loosely wrapped. Around the loaf itself was a band of card, printed in gold and red, with the Barilla logo, a sun symbol and the name for the new bread, Panem, which means bread in Latin. The double wrapping was a reminder of how a traditional baker would have sold bread, while allowing the client to appreciate the product. The name Panem, with its sonorous overtones of the Mass and classical literature, reflected the serious importance of the product. It now enjoys a significant market share.

Guiltless Gourmet Taco Chips

FOOD

DESIGNER Cindy Goldman Design and in-house team
CLIENT Guiltless Gourmet, Austin, Texas, USA

5 — 38/39

TYPE	Extension of existing range plus shipper pack (see over)
PRODUCT	Food snack
MATERIAL	Printed heat-sealed foil (chips), glass jars (dips)
SIZE	7oz and 12oz (198gm and 300gm) pack
TIME	5 weeks
ELEMENTS	Barcode, nutritional panel, ingredient declaration, product name and company logo, name and address
MARKET	Female, head of household, upper to middle income level, health-conscious adults aged 19–29
SUPPORT	In-store danglers, general advertising
KEY WORDS	Healthy, better for you, snack food

DESIGN BRIEF

Guiltless Gourmet, by their own account, was founded in 1989 'by a group of slightly-overweight tortilla chip devotees who wanted to diet without giving up their favorite snack food'. Today it is one of the fastest-growing privately-held companies in the USA, and their tortilla chips are the tenth best selling brand in the country. They have also created a range of fat-free dips and salsas, and they introduced a range of potato crisps in the summer of 1996. The key to their approach to a low-fat chip was to bake rather than fry the product, and so 'baked not fried' is the key strapline on the chips. This is reinforced by a gold button reading 'no oil added' or '1 gram fat', while 'fat free' is used on the dips. The Guiltless Gourmet logo used capitals for the first word to reinforce the health message which is the basic selling proposition of the product range. The products also carry endorsements from heart associations and the 'State of the Heart' trademark.

' We wanted to create a spectrum of color that nobody else was using, radically different from what everybody else had on the shelf. '

DESIGN According to the package designer Cindy Goldman, 'we needed packaging that would catch the eye in the third of a second that you have to catch the consumer's eye'.

Two technical innovations were used to achieve this in the final design: high barrier film, to ensure a better look to the product and better taste preservation, together with printing the packs in Pantone colors rather than from a four-color process. This achieves a higher degree of color saturation.

Key Factors

DESIGN Snack products are fun: so the design has to have immediate and bright visual appeal to draw attention. — P.38

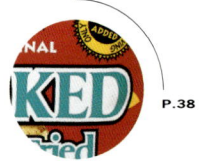

MATERIALS The choice of materials both attracts the consumer and increases shelf life. — P.40

SUPPORT To create a visual link across two product ranges, color ways and logos must mix well. — P.41

Strong colors are printed on foil for added visual impact while the foil guarantees the freshness of the product.

Guiltless Gourmet – design development

Designer	Mark Green and in-house team
Cilent	Guiltless Gourmet, Austin, Texas, USA
Product	Food snack shipper box
Material	Printed card

Design Development

'Your packaging is like a billboard sitting next to a busy freeway. It is only successful if prospective customers can get the message immediately. GG's Taco chips are baked not fried, and were the first of their kind. Even today, consumers have to understand quickly and easily this point of difference.'

A shipper box has a double function. It is used when the product is transported from the distribution center to the store, and can then be used as a display unit in itself. In retailing, shelf space and floor space are at a premium. Space for your product means less space for a competing one, after all. But retailers are unwilling either to invest in special display areas themselves, or to promise permanent space to manufacturers' displays. This shipper/display pack for chips or for salsa is an example of gaining space without asking for the retailer to make a long-term commitment. The outer carton serves as the delivery package for three standard boxes of packets of chips. In-store, the front panel is torn off, displaying the products on their own 'shelves'. Part of the front panel is slid into the top to make a header card.

Materials

The advantage of the shipper box system is that it is low cost at both ends. The manufacturer has to overprint and perforate the outer carton, but would be involved in the costs of the carton anyway, to ship the product. The retailer can decide either to display the opened shipper, or simply stock the shelves from it. After use, the retailer can either refill the display or throw the carton away.

The design principle for the shipper was 'to create maximum impact on the floor while controlling costs'. According to Jim Bream of Guiltless Gourmet, 'it proved to be exceptionally affordable for a fully-functional pre-packed floor display'.

SUPPORT

The design philosophy of the chips carries over into the designs for salsa (facing), using strong colors, color coding and an informal, relaxed logo. Shelf danglers are used to link the products in-store: there is also a shipper box system for the dips and sauces.

Your packaging is like a **billboard** *sitting next to a* **busy freeway.**

The use of color coding to distinguish different flavors is carried into the use of different colored type in the logos.

Detail shows how the foil back and screen print reinforces the color appeal of the pack.

The shelf dangler provides a visual synopsis of the product range at point of sale.

Seeing Through Glass

Drink

The concept of design equities is fundamental to an understanding of packaging design. The term means those aspects of a product which represent its public perception: a name is the commonest example, either of a product or a company. A logo or a special pack shape are other examples. These can be protected by trademark and copyright legislation, but their value goes considerably beyond this. An equity can be a color linked in the public imagination to a product (imagine Coca-Cola in a green tin!), or an advertising slogan, such as 'Guinness, pure genius'. A designer looking at a redesign needs a clear perception of these equities, and any redesign work, if it is to be successful, needs to support and strengthen them.

Sometimes the company's own equities are the most important: this particularly applies, as we saw with Safeways, to own-brand products, where the full status of the company stands behind the individual product. Sometimes the product has its own equity: many of the most famous names in whisky, for example, are owned by one company, International Distillers, but few whisky drinkers are influenced in their choice by this fact.

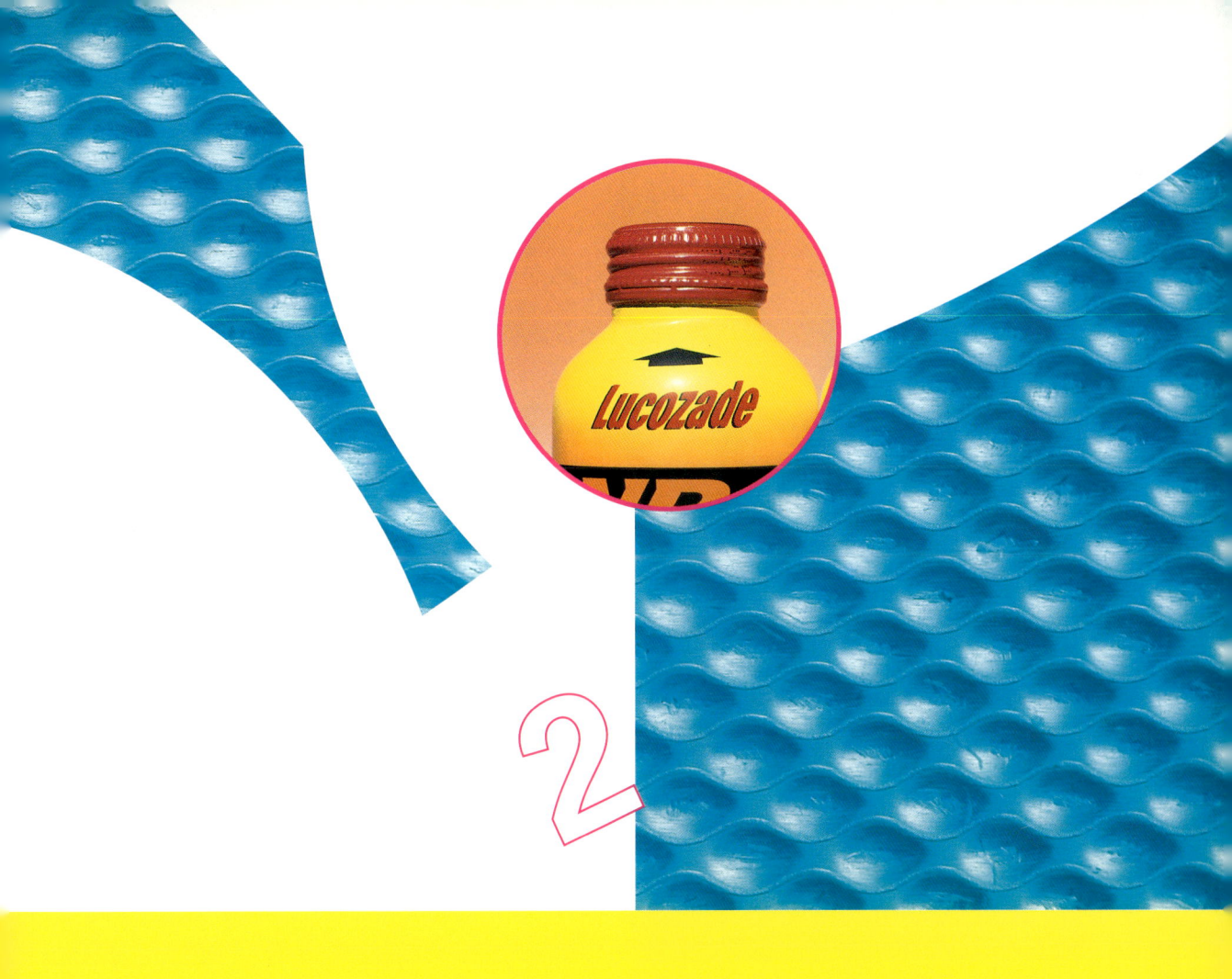

A designer creating a new product, particularly in a crowded market sector, needs to plan ahead to identify the potential equities in the new product, and to link this analysis to extensive market research before design and to the development of the marketing and launch plan. After the launch, the success of the product needs to be monitored and the design evolved to support the emerging equities.

In this selection of designs for the drinks market, we have tried to identify products that are either competing with existing brands in creating new equities (such as the Adelphi Distillery's malt whiskies), or maintaining established equities in a niche market (such as Leinenkugel's beers and ales), or developing existing equities through a special product (as with Smirnoff's vodka gift-pack to celebrate the reopening of their Moscow brewery). Other studies look at the use of materials (with Sainsbury's red wine) or unique packaging solutions (in Lucozade Sport); but in these cases also the equities of the brand and its development also play a key role.

DRINKS

Sainsbury's Vin Rouge

DESIGNER In-house design commission
CLIENT Sainsbury's plc, London, UK

1 44/45

TYPE	Redesign of existing packs
PRODUCT	Red table wine
MATERIAL	Polypropylene bottle, paper label
SIZE	1.5 litre & 75ml bottles
TIME	4 months
ELEMENTS	Barcode and company logo
MARKET	General adult
SUPPORT	In-house magazine advertising
KEY WORDS	Enjoyable, bright, economical

Design Brief

Red wine is traditionally bottled in green or brown glass, since direct light can, over time, cause chemical changes in the color and, in turn, the taste of the wine (white wine, which does not have such a problem, is more commonly bottled in clear glass). When putting inexpensive wine for immediate drinking in plastic bottles the same conventions were at first followed. The resulting green translucent plastic has a matt surface, unfortunately emphasizing the cheapness of the product.

Sainsbury's, the UK supermarket chain, have refined this process for their French red wine bottles. The change is simple but significant: green translucent plastic is out, and a clear plastic used instead. This plastic has a very high reflective surface, so that it, and the wine within, literally sparkle. The change lifts the product's visibility on the shelf, and makes it an attractive purchase rather than a purely economical one.

At first, traditional bottle shapes were used. But as the techniques of injection moulding do not relate to the blown glass tradition of bottle-making, it soon became clear that a larger bottle (containing 1.5 litres, or just over two standard bottles) was more stable, and could be transported better, if the long neck was abandoned and a square rather than a round profile adopted. By including ribs of indentations the structural mechanics of the bottle function better, allowing a thinner and so lighter plastic envelope.

The label uses shades of red, turquoise and ochre with a simple motif of a bottle that sits on the corner edge, and the product name (set in Fenice) printed in dark red on white. The choice of a busy face like Fenice contrasts with the technical information, given in a narrow sans serif. The Sainsbury's name is in the house style.

Key Factors

MATERIALS

The choice of a very clear material communicates the product in the best possible way.

P.44

MANUFACTURE

The manufacturing requirements of a plastic bottle are used to advantage, with the ribs acting as a measure.

P.44

Maintaining the same label visuals across both glass and plastic bottles reassures the buyer as to quality.

Lucozade Sport and Energy

DRINK

DESIGNER In house team
CLIENT Smith Kline Beecham, Brentford, UK

2 | 46/47

TYPE	Repositioning design for new market
PRODUCT	Health sports drink
MATERIAL	Glass bottle, paper or plastic label
SIZE	15oz (330ml) pack
TIME	Six months
ELEMENTS	Barcode and company logo, nutritional information, best-by date
MARKET	General youth/adult sports and health
SUPPORT	Television and poster advertising
KEY WORDS	Healthy, fresh, quality product

DESIGN BRIEF

The traditional bottle had strong equities, but fixed in one market. The new packaging had to extend the market.

Lucozade was invented in the 19th century by a dispensing chemist in Newcastle-upon-Tyne, who created the glucose-based drink as an aid to convalescence. It has since the 1930s been developed and marketed by the healthcare company Beecham's, now part of Smith Kline Beecham. The original packaging for this effervescent drink was a glass bottle wrapped in orange-tinted cellophane. Its appearance on countless childhood bedside tables was a sure sign that illness was on the wane!

' We develop products that have a health advantage as a premium element, and our design policy reflects this. '

Key Factors

DESIGN
The rapid development of a 'health drink' market for sport and fitness demanded new design solutions.
P.48

MATERIALS
New materials for labelling created new opportunities for more integrated design.
P.49

DEVELOPMENT
Alternative package solutions carried the new strategy into further retail and market options.
P.50

A transparent 'body label' retains visual equities but makes a stronger product statement across the whole bottle.

Lucozade Sport and Energy

The first development was a clear plastic bottle. The easy to grip, contoured surface suggested a sports context.

Type of Design

To launch the newly positioned product a new taste – Tropical Orange – was launched in parallel, and this was followed by a larger bottle size, in polyethylene, as an alternative. These bottles had simple paper labels. The last addition to the bottle range is Lucozade Energy in a new 300ml glass bottle with a transparent label covering the whole bottle. 'We could use the whole of the bottle surface to promote the product,' according to Geoff Giles. 'Instead of thinking of the label as something stuck on the bottle, we looked at the product as a whole.'

Total labelling and unusual bottle shapes are also adopted for the latest product, Lucozade NRG.

We develop products that have a **health advantage** as a premium element and our **design policy** reflects this.

MATERIALS

While the cellophane wrapped bottle is still in use in some overseas markets where the product is still identified as a health aid, in the 1980s Beecham decided to reposition the product as a drink for the growing sports market. According to Geoff Giles, category head of the packaging development department, this change needed to retain the intrinsic value of the product, while presenting it to a new market. The first solution was a short ribbed bottle. The ribs make the bottle easier to hold, as well as recalling the folds of cellophane. The bottle shape was also deliberately different from other products in the field, with a wide neck.

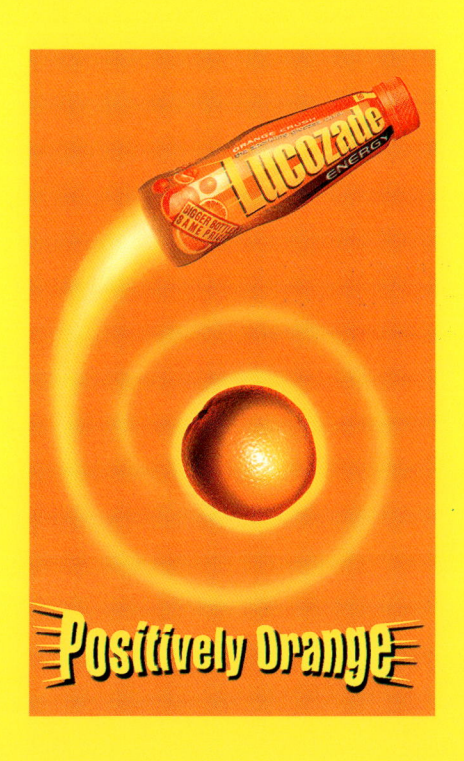

Advertising support stressed the 'energy' theme and the contemporary shape of the bottle.

DRINK

Lucozade Sport Pouch – design development

2
50/51

Designer	In-house team
Cilent	Smith Kline Beecham, Brenford, UK
Product	Health sports drink
Material	Foil pouch using the Guala system

Design Development

The high-tech cap, Olympic-style logos and angled type position suggest a serious approach to sport and fitness.

To emphasize the sports connection for Lucozade, Geoff Giles and his colleagues looked at alternative packagings. They selected the Gualapak system, developed in Japan and marketed in Europe by Guala in Italy. The technical aspects of how the container is filled and sealed (as it is a food product) were an important consideration. The containers are hot-filled and then chilled, so the packaging has to allow for changes in volume of the liquid. With a glass bottle there is no problem, but polyethylene bottles, for example, need sides of variable thickness to allow for expansion and contraction. The top of the container also needs a firm seal, both to retain freshness and deter tampering. The Guala pack, in metallic foil, has an integral straw and a spigot seal.

The back of the pack provides space for technical and nutritional information, barcode and company logos, and promotional offers.

Research tells us that the metal foil is seen as a premium and quality product.

SUPPORT

'We considered developing our own system, but the time that would be involved in prototyping and setting up manufacturing did not warrant it. Guala normally only issue one licence per country, so the packaging would be identified uniquely as ours in the main market.'

'Research tells us that the metal foil is seen as a premium and quality material, so the fact that the contents are not visible is not a problem. We are targeting the same market with the bottle and pouch products, as well. The soft pouch offers an alternative which can be slipped into a sports bag or large pocket.'

Market acceptance of the new pouch has been good: the next stage will be to develop shelving systems for instore displays.

The flexible qualities of the new Guala pack key into an open, sports lifestyle.

DRINK

Adelphi Whisky

DESIGNER Graham Scott & McIlroy Coates, Edinburgh, Scotland
CLIENT Adelphi Distillery, Edinburgh, Scotland

3   52/53

'In a saturated market for a connoisseur's product, clear, elemental design has a major role to play.'

TYPE	New design and development for new markets
PRODUCT	Malt and fine Scotch whiskies
MATERIAL	Clear glass bottle, printed paper label, recyclable card box
SIZE	75ml bottle
TIME	4 months
ELEMENTS	Standard listing of strength and quantity of spirits
MARKET	Connoisseur drinkers
SUPPORT	Direct mail leaflets
KEY WORDS	Exclusive, direct, discreet

DESIGN BRIEF

Whisky is distilled from grain. Traditionally in Scotland small individual distilleries each produce in pot stills an individual spirit, its taste and color distinct from others because of local water and other factors. Some of these whiskies are sold as single malt whiskies, others are mixed with grain whiskies (made in patent or Coffey stills) into blended whiskies which are sold under generic names. Blending different malts with grain whiskies produces the range of colors and flavors that are one of the strengths of the Scotch whisky market.

Key Factors

DEVELOPMENT
The design approach must be sufficiently flexible to carry over into new applications without losing core direction. 
P.52

MARKETING
Direct mail products do not need to compete for shelf space, but must make an immediate statement of their position. 
P.53

ELEMENTS
In selling to a sophisticated market, design elements can be used to make important points of detail. 
P.54

SUPPORT
Sophisticated products need sophisticated information, which in turn develops further sales. 
P.54

The design is minimal so that the consumer will be able to 'read' the bottles, using the product's natural color.

Adelphi are whisky brokers, not independent producers. They purchase casks of whisky from single malt producers, and sell them by direct mail and catalog to connoisseurs of fine Scotch. Since they only bottle a limited quantity, the product is effectively a limited edition in each case. The customers are generally well-informed about whisky, and interested in the pure flavor of the 'water of life' rather than a hard sell in an overblown package.

The logo uses both type and freehand lettering to balance tradition, status and a contemporary approach.

ADELPHI
Distillery
LIMITED
EDINBURGH SCOTLAND
1826

Adelphi Whisky – design development

The label is positioned deliberately two-thirds down the bottle to allow the color of the whisky to show through.

Support

The external packaging does not need to compete for shelf space, given the direct mail sales method. So it is made from recycled cardboard with a paper label. The result is elegant: sufficiently tasteful to offer the whisky as a gift, and not so ostentatious as to deter a connoisseur. Each bottle is accompanied by a well-printed, simple booklet describing the whisky and the distillery. These booklets build into a library about Scotch.

Elements

Not only does whisky not discolor with age, but the color of the malt spirit is an indication as to where and when it was produced. A plain but elegant clear glass bottle with a simple label was the obvious solution. The label carries a roundel which is in turn color-coded to show the area of Scotland where the whisky was distilled.

Designer	Graham Scott & McILroy Coates, Edinburgh, Scotland
Cilent	Adelphi Distillery, Edinburgh, Scotland
Product	Malt and fine Scotch whiskies
Material	Clear glass bottle, printed paper label, recyclable card box

Design Development

By retaining the key elements, the design formula can be extended to new products within the same family, even if for different markets.

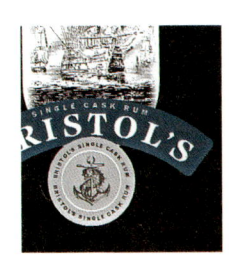

The success of Adelphi's range of specialized single malt whiskies in the UK and the USA led to enquiries from agents in Russia and Japan for a similar range of products for their markets. However, single malt whiskies are not as well known there, and the local preference is for special blended whiskies. (In the retail sector, this is shown by the fact that a top-range blended whisky such as Johnny Walker Black Label sells for more than single malt whiskies.)

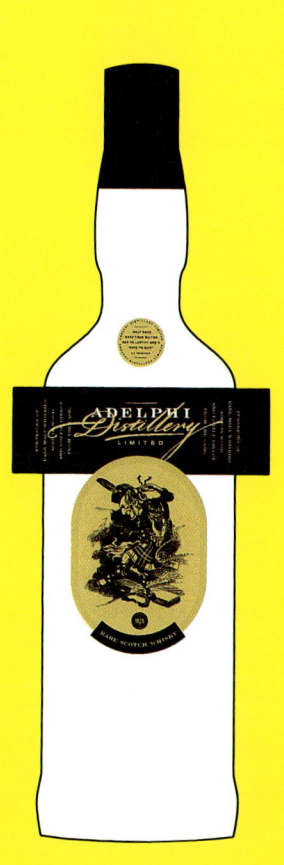

Development

Graham Scott designed a series of labels for a blended whisky, to be marketed under the Adelphi name exclusively in Russia. These retain the formal values of the main range, as well as the bottle shape. The supporting material is in a different color, and is rather more general, not to say whimsical in tone. This reflects the different level of information about whisky in the Russian market. A similar development is planned with vintage port.

DRINKS

Smirnoff Vodka

DESIGNER Raymond Loewy International, London, UK
CLIENT International Distillers, London, UK

4 56/57

TYPE	New presentation pack design
PRODUCT	Speciality vodka
MATERIAL	Plastic, cellophane, card
SIZE	75ml bottle
TIME	4 months
ELEMENTS	Boxes for bottle and gift glass
MARKET	General adult
SUPPORT	Specialist magazines
KEY WORDS	Exclusivity, luxury, facility

DESIGN BRIEF

Gift packs for luxury goods are a key part of a modern marketing strategy, and are often put into place at festive times of the year such as Christmas and New Year. However, since presentation packs take up more shelf space than simple bottles or boxes, there needs to be a sufficient price premium and visual incentive to warrant them. There is also the physical problem of ensuring the pack travels well, so retaining its added value up to presentation.

The Smirnoff logo is an established equity: by varying the color, however, it can differentiate and unite products.

'*The Smirnoff design was primarily a celebration, but required an exacting approach to the market in order to succeed as well as it did.*'

Key Factors

DESIGN
A gift product has to have visible added value both in content and presentation.

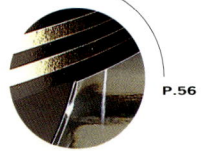

P.56

MARKET
The marketplace has to be understood fully if special products are to achieve success.

P.58

ELEMENTS
A premium product has to travel safely, and remain good in display. So a structural packing solution can often be appropriate.

P.59

SMIRNOFF

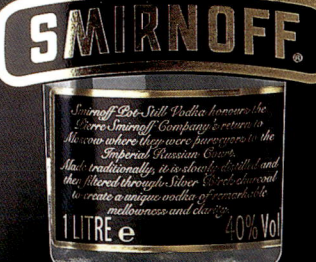

Ste Pierre Smirnoff Fils
TRADITIONAL
YELLOW
RUSSIAN VODKA
дистиллировано в Москве
DISTILLED & BOTTLED IN MOSCOW

SMIRNOFF

1 LITRE ℮   40% Vol

IMPORTED
1 LITRE ℮ 40% Vol

2
Flute Glasses

Smirnoff Vodka – design development

PRODUCT	Speciality vodka
MATERIAL	Plastic, cellophane, card
SIZE	75ml bottle
TIME	4 months
TYPE	New presentation pack design
ELEMENTS	Boxes for bottle and gift glass
MARKET	General adult
SUPPORT	Specialist magazines
KEY WORDS	Exclusivity, luxury, facility

Design Development

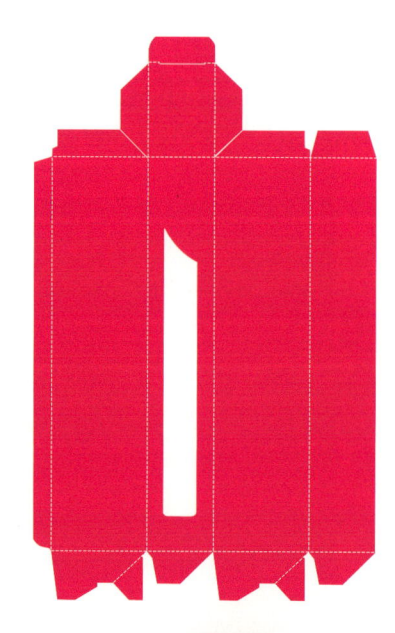

When Smirnoff consulted Loewy International about a pack for their premium black vodka to celebrate the reopening of their Moscow distillery, the elements mentioned on page 56 came up for consideration. Earlier closed presentation packs in card had been damaged by customers who wanted to see the contents, so a see-through pack was required, with the bottle behind clear acetate for security. As a premium, it was proposed to pack the bottle with two special glasses, increasing its footprint on the shelf. However, the retailer does not want to lose shelf-space unless the premium is considerable. Loewy designed a pack that increased the footprint by only 50%, rather than by a full bottle space. They added a further option: by dividing the pack into two elements the retail opportunities were increased: you could buy just the bottle, or the two elements together. The design of the two units was co-ordinated so that they would work independently or together.

The development of the structural pack required considerable manufacturing expertise.

360mm

Market

Loewy's response to the brief was not only an elegant design with clever technical aspects, it also shows a keen awareness of the retailer's attitudes to premium products. The success of the new packaging is reflected in sales of over 600,000 units worldwide.

The structure's base holds the bottle in the correct position, while the liner is foiled in silver so as to increase the luminosity of the package.

ELEMENTS

The drawback to a clear pack was that the bottle could rotate in transit, leaving the customer looking at the bar code. For Smirnoff, Loewy created an innovative structural packaging using a vacuum formed plastic anchor structure to hold the bottle in place and ensure the label was also centrally visible. This board liner was coated in silver to create an aura of light around the bottle, a feature used in the matching acetate pack for the two shot glasses.

DRINKS

Virgin Cola Pammy

DESIGNER Ashley Stockwell, Virgin Trading, and Start Design, London

CLIENT Virgin Cola, London, UK

5 60/61

'*The Pammy bottle will make other colas seem square.*'

TYPE	New design
PRODUCT	Sparkling soft drink
MATERIAL	Printed paper label on polythene bottle
SIZE	500ml bottles
TIME	8 months
ELEMENTS	Legal information and company name and address
MARKET	General youth market
SUPPORT	In-store displays, poster advertising
KEY WORDS	Shapely, sexy, trendy

DESIGN BRIEF

The Virgin group of companies, founded and owned by Richard Branson, began in the record business and has since diversified into the airline business, pensions and drinks, marketing their own vodka and cola. The next development in line, it is said, is to be blue jeans. The company sees informed but informal, hip youth, both male and female, as its main target. Their top-down leadership makes them extremely flexible, and they have taken on some of the stiffest competition in their markets, notably British Airways on the transtlantic routes from London to New York.

MARKET In moving into the cola market, they were taking on two very big players, Coke and Pepsi. Coca-Cola have set the benchmarks for bottle design since 1916, when their distinctive profile bottle was introduced. Eighty years later Virgin were looking for a way to redefine the cola bottle. They hit on linking the product to the star of the immensely popular Baywatch television series, Pamela Anderson. The bottle is endorsed by her and the shape of the bottle is based on her vital statistics. A limited edition 375ml glass bottle carries her signature.

By seeing product design as communication, Virgin were able to link their product with a popular icon in the target market.

Key Factors

MARKET

A link with an established figure is a key way of establishing product endorsement.

MATERIAL

A new manufacturing process should be judged in the context of its total design effectiveness.

P.60   P.60

The bottle required new moulds and careful quality control to meet the design, but this cost was justified by the market success of the new cola.

Quady's Winery

DRINKS

DESIGNER McCoy Design, Los Angeles, CA, USA
CLIENT Quady's Winery, Montara, CA, USA

PRODUCT	Dessert wines
MATERIAL	Printed paper labels
SIZE	75ml bottles
TIME	Various
TYPE	New designs
ELEMENTS	Legal information and company name and address
MARKET	General
SUPPORT	In-store displays
KEY WORDS	Sophisticated, unusual,

Design Brief

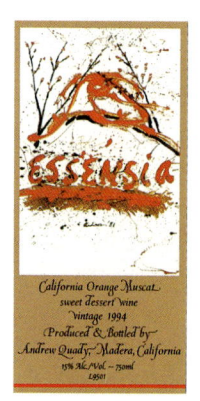

Quady's Winery in California is a new addition to the wines of the New World. It was started by Andy and Laurel Quady in 1980 with the launch of Essensia, an Orange Muscat dessert wine. The Winery now has a range of seven dessert wines and a liqueur, Spirit of Elysium, double-distilled from black Muscat grapes. A careful supervision of the fermentation process has resulted in a group of wines that emphasize flavor and aroma rather than alcohol content.

The choice of dessert wines came from Andy Quady's own interest in making port (and perhaps from a realization that table wines in California now require almost industrial levels of production to achieve economic viability). Dessert wines are special purchases, for occasions such as a dinner party or celebration, rather than for everyday drinking. Despite a higher price tag than table wines, ensuring a larger return for the retailer, they need to compete for shelf space against faster-moving lines. The presentation of the wines, both in terms of labelling, point of sale material and advertising, had to be carefully targeted to achieve success.

' Selling Essensia is like selling dessert. It must be suggested. '

Key Factors

DESIGN
Introducing a new product into an established market requires a sophisticated design approach.
P.62

MARKET
Premium and gift packages can be used to support a new product or brand in a tight retail environment.
P.64

DEVELOPMENT
As the new brand is established, design must be used to re-inforce the individual quality of new products.
P.65

QUADY • QUADY
ELYSIUM
CALIFORNIA
BLACK MUSCAT
SWEET DESSERT WINE
VINTAGE 1995
PRODUCED AND
BOTTLED BY
ANDREW QUADY
MADERA, CA
15% Alc./Vol. ~ 375ml
L9601
QUADY • QUAD
ESSENSIA
California Orange Muscat
sweet dessert wine
Vintage 1995
Produced & Bottled by
Andrew Quady, Madera, California
15% Alc./Vol. ~ 375ml
L9601

Quady's Winery – design development

Artistic and individual labels can convey both individual quality and a brand approach.

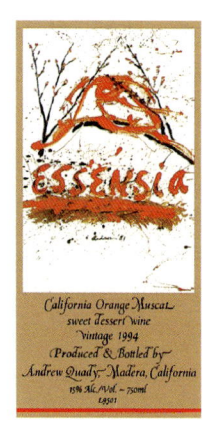

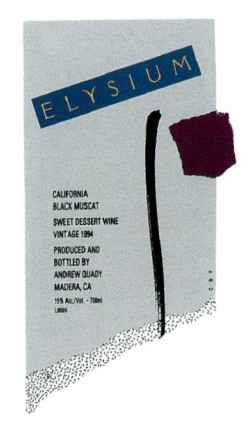

Market

'Selling Essensia is like selling dessert,' according to a brochure for retailers. 'It must be suggested. Introduce our wines as a dessert substitute, accompaniment or for after dinner. Many appreciate a new experience: something different and really good.' The labels for Essensia and the accompanying dessert wines, Electra and Elysium, follow the same strategy of choice and distinction. For Essensia and Electra gold-bordered labels are used, decorated with commissioned paintings, each incorporating the name. These suggest that the wines are right for the young and aware who enjoy the challenge of contemporary art. Elysium uses a more abstract design, together with a torn label to suggest, again, an art connection.

Support

This connection is not overplayed (even though their older vintage port bottle uses a detail from an Impressionist painting) but acts as a visual mirror to both the delicate and sweet tastes within, and the sensibilities of potential purchasers.

Through the price premium they offer the retailer, gift packs and multiples are an important part of developing retail space for the product,

Selling Essensia is like selling dessert. It must be suggested.

DESIGNER	McCoy Design, Los Angeles, CA, USA
CLIENT	Quady's Winery, Montara, CA, USA
PRODUCT	Dessert wine
MATERIAL	Silk-screened bottle
SIZE	75ml bottle

DESIGN DEVELOPMENT

To achieve credibility as an independent and innovative producer of dessert wines, Andy Quady realized he had to have his own port wine, even though the very use of the name port was becoming increasingly restricted. A trip to Portugal revealed that one of the traditional port grapes was grown in California under the name Valdepeñas, and so Quady decided to start testing these. The result was a pair of wines with a distinctive port character.

The choice of a name for these was partly an inspiration, partly a joke, and partly a declaration of independence. Where could be further to the west of Portugal (and so to the right on the compass card) than California? Starboard it had to be! But having made that choice, Quady decided to follow it by adopting a completely different approach for labelling. No illustrations, just a logo looking something between a polar star and a gear-wheel, a somewhat stark and futurist image. No paper label either, rather the design is silk-screened, in red or white, directly onto the glass. This sets it apart from the other products in the brand, emphasizing difference, but also harks back, in a wholly new way typographically, to a tradition in bottling port of screened rather than printed labels.

This careful balance between the familiar and the unexpected, especially in the Starboard Batch 88 bottle, is the result of a firm overall view by the client combined with graphic flair on the part of the designer, John Coy of Coy Associates in Los Angeles.

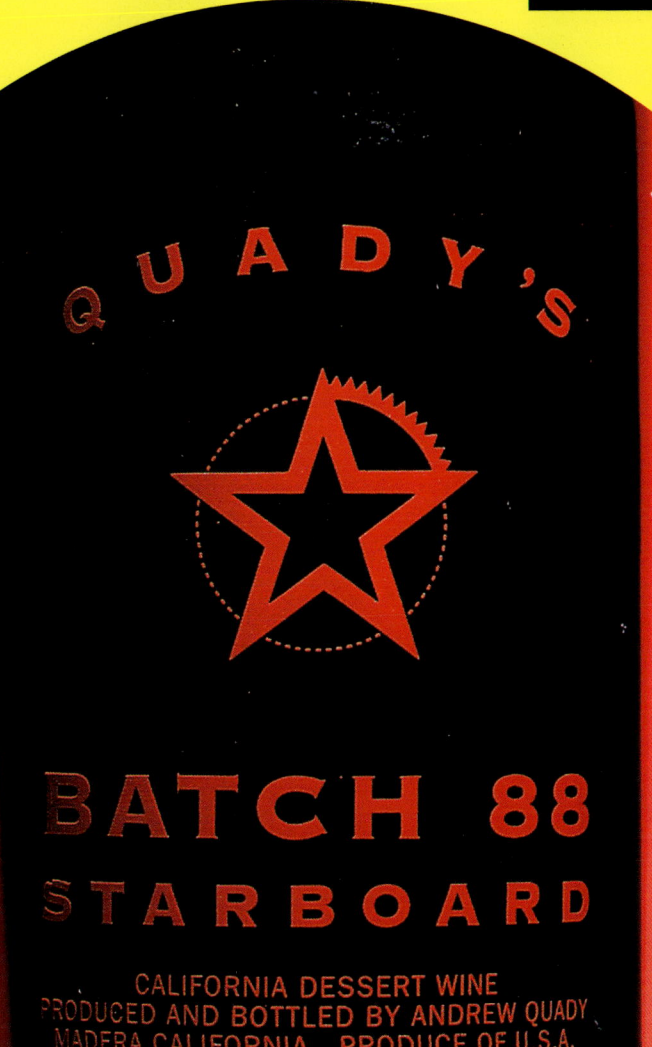

DRINKS

Leinenkugel's Beers

DESIGNER Design Partners, Racine, WI, USA
CLIENT Jacob Leinenkugel's Brewing Co. Inc., Chippewa Falls, WI, USA

7 | 66/67

TYPE	Redesign of existing packs plus new brands
PRODUCT	Speciality beers
MATERIAL	Board six-pack boxes, glass bottles, printed paper labels
SIZE	Standard 12oz bottle
TIME	2–4 months
ELEMENTS	Barcode, alcohol content and company logo
MARKET	Male and female adults in the 25–50 age
SUPPORT	Local television and press advertising
KEY WORDS	Traditional, hand-crafted, original

DESIGN BRIEF

Jacob Leinenkugel emigrated from Bavaria to the north-west of the USA in the 1860s, and set up his brewery in Chippewa Falls in the Wisconsin North Woods in 1867. Five generations of Leinenkugels have carried on the tradition, although the company is now part of the large Miller Beer Company. Three brothers, Dick, Jake and John, still run the brewery. They concentrate on producing small batches of hand-crafted ales, beers and lagers, often seasonally related. These are intended for a relatively sophisticated market who are interested in beers with unusual flavors, such as the Honey Weiss, a white lager beer flavored with local honey.

Key Factors

MARKET
Products in small batches within a brand need maximum on-shelf support, even with above-the-line support.
P.66

DESIGN
A generic approach to design allows fixed elements to be maintained while giving each product run its own personality.
P.67

SUPPORT
A varied brand needs to create a sense of consumer identity to build support. New technologies are one key to this.
P.71

The richly decorated box conveys brand values, as well as retaining space for a relatively small product in a busy market.

DESIGN For the designers, the challenge was to produce a design approach that would allow each product to express its individuality while remaining part of the overall brand. Since the Leinenkugel's signature has been a standard element in the design for some time, it was made into a main feature, both on bottles and six-packs. Since the company's family history is a key part of the product's appeal, a Victorian look with type on straps and ribbons and ornate old typefaces has been deliberately adopted. Rich gold and burgundy edgings are supported by specially-commissioned illustrations.

Here's another one from jake and the boys!

Leinenkugel's Beers – design development

DESIGNER	Design Partners, Racine, WI, USA
CLIENT	Jacob Leinenkugel's Brewing Co. Inc., Chippewa Falls, WI, USA
PRODUCT	Speciality beer with autumn fruits
MATERIAL	Board six-pack boxes, glass bottles, printed paper labels

DESIGN DEVELOPMENT

The tradition of brewing beer with added fruit flavors has long been established in Germany, Holland and Belgium, but is less well-known elsewhere. In the very competitive market for speciality beers, fruit beers offer a good route to new products. Because Leinenkugel's is linked to speciality, seasonal beers are not a disadvantage in production terms. In design terms, the fixed elements (signature and roundel, for example) have an additional importance in reminding the client that 'here is another one from Jake and the boys!'.

Individual labels remain individual, but the roundel and ribbon are consistent elements.

MATERIALS

So while each label and six-pack has an individual design, the linking factors are equally important. From a cost point of view, all the six-packs use the same cutter plan (see facing page). While it is complex, it is economical in card use, and is a neat example of paper engineering: the use of folds to strengthen key points in the design, particularly around the handle. At the same time the assembled design also packs flat for transport from printer to brewery. (Two clever touches are how the roundel fits into the handle when the pack is flat, and the way that the hook that locks the box open is protected within the box when it is flat, reducing the risk of damage before assembly.)

The deliberately ornate lettering and gilding act as a reminder of the company history, as well as grabbing the eye.

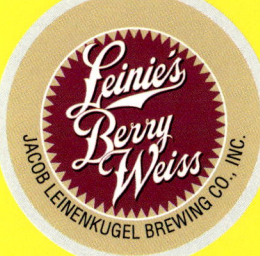

Colors and illustration are repeated across box and bottle so that both can be identified whether on the shelf in packs or displayed singly.

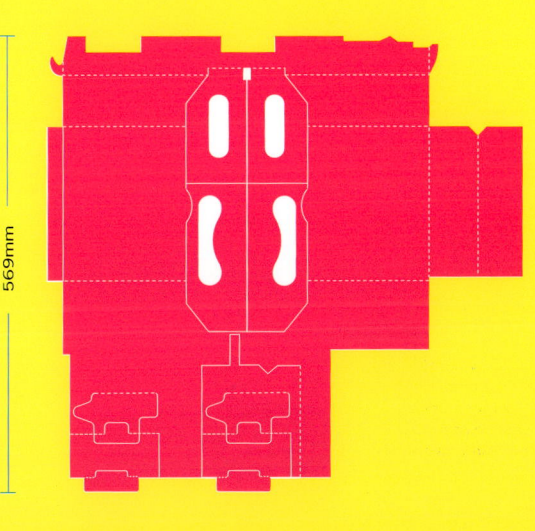

Design

Design Partners' work for Leinenkugels is a good example of teamwork within the office and co-operation with the client. The designers need an understanding of the brewer's heritage and its important role in their marketing. (When Leinenkugel's was acquired by the Miller Brewing Company in September 1995, the press release stressed both the history of the company and the fact that day-to-day running would remain in the family's hands, for example.) As to teamwork, each design is created by a group of at least five: Rick Petroske, from Miller, as creative director, R. David Foster as account executive from Design Partners, a designer, and one or more illustrators and production artists.

The flat pack is standard across all products in the brand, so reducing manufacturing costs.

Leinenkugel's Beers – design development

DRINKS

There's plenty to see at Leinie Lodge: the tap room, a gift shop, Jake's den, where you can read about his travels, or just a front porch to sit and sip!

DESIGNER	Design Partners, Racine, WI, USA
CILENT	Jacob Leinenkugel's Brewing Co. Inc., Chippewa Falls, WI, USA
PRODUCT	Speciality beer with autumn fruits
MATERIAL	Board 12-pack boxes, glass bottles, printed paper labels

LENNIE LODGE SAMPLER PACK

Leinie Lodge is a virtual place: it is the home page of Leinenkugel's site on the World Wide Web. The brewery is on the horizon: click on it and you can see the process at work. Within the lodge itself there is a tap room (only for virtual drinking, alas), a souvenir shop, a member's clubroom, and Jake's den, where Jacob Leinenkugel offers his thoughts about beer and stories about his travels. The atmosphere is one of comfortable bonhomie, the interior decoration typical of a country cabin, with rough-cast walls, log fires and comfortable furniture. The design for the lodge first appeared on the label and pack for Berry Weiss.

Leinenkugel's success depends on their maintaining a consistent overall approach to their product, and combining this with a linked series of designs, promotions and products.

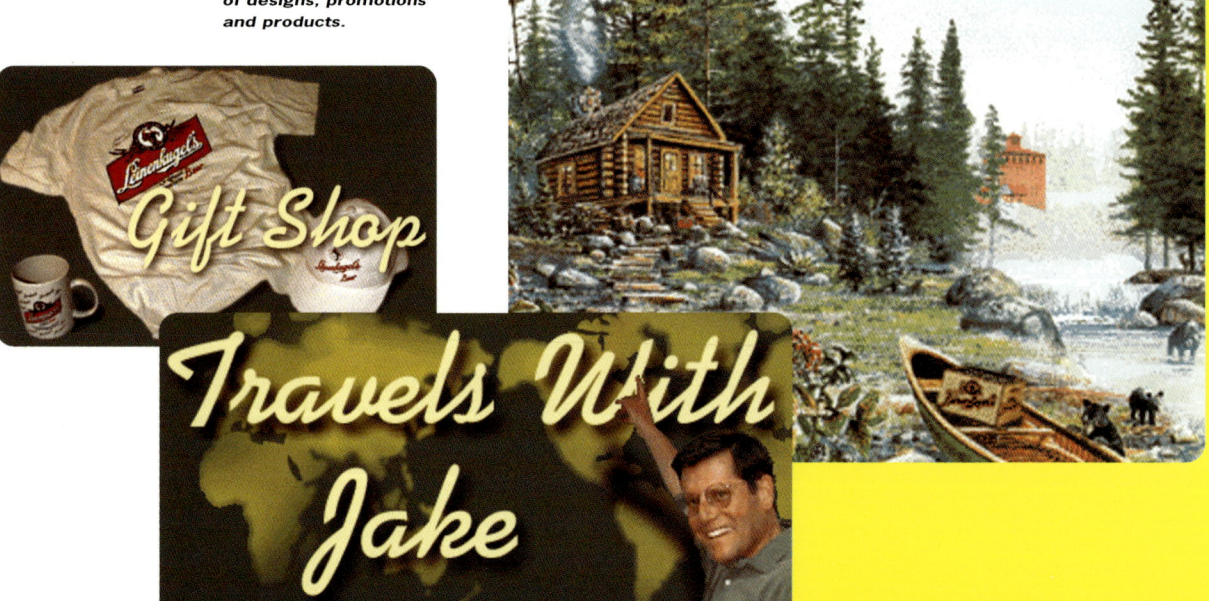

SUPPORT

The use of a Web site to promote a product, brand or company is becoming increasingly necessary as more and more potential consumers come to discover the Web and use it both for research and leisure. The Leinenkugel's site is well planned, offering a balance of product information and special offers (some for members only). More importantly, it offers positive support to the Leinenkugel business ethic, by giving the history of the firm and its aspirations a visual form.

The Leinie Lodge pack builds on this approach. It is a special promotional 12-pack box, offering a selection of different beers as a series of samples of the Leinenkugel range. It shows how a design initiative, if it is correctly planned, can move from a physical design to a new medium and then in turn spark a new design and marketing idea.

3 Leisure

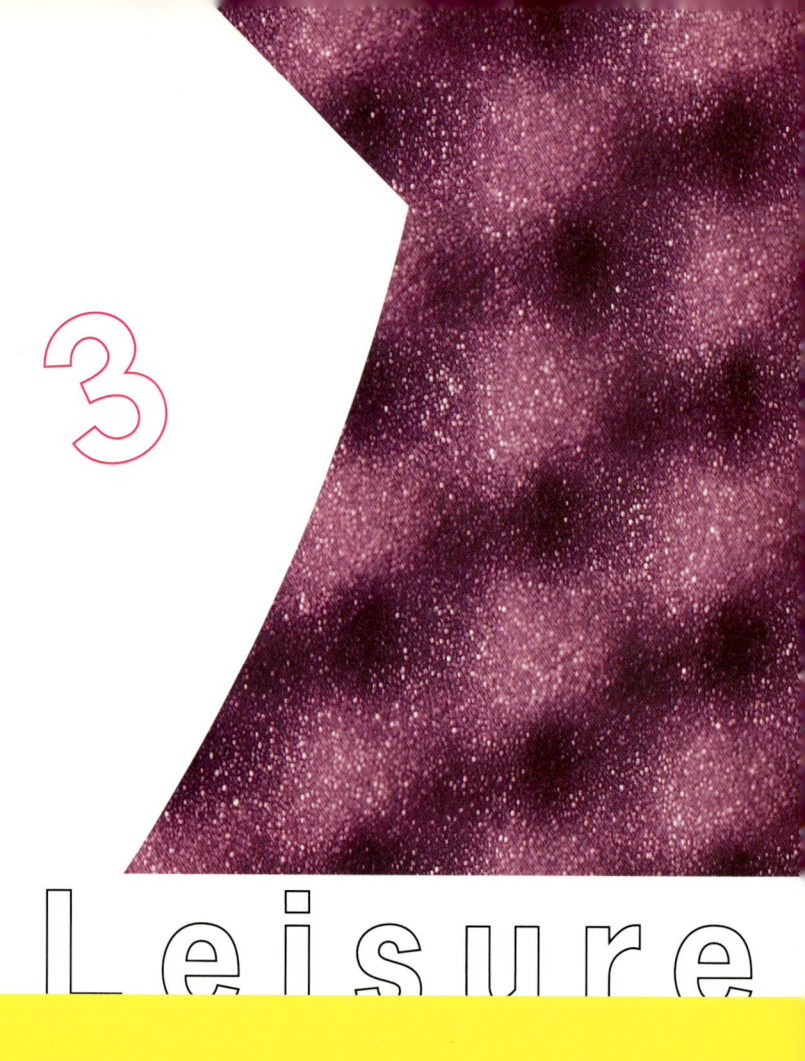

Wrapping the Gift

The shopping bag has become a fashion accessory: even, some say, a status symbol, citing the shoppers in cut-price stores who carefully place their purchases in a Harrods or Bloomingdales bag! Din's careful work for Nicole Fahri shows just how important this design area still is. The bag needs to be integrated with the client's perception of the store as a whole, the same problem faced by David Chan in designing a range of packages for the Orient Tea Room in Hong Kong.

Toys are premier gift items, bought by parents and children, and by grandparents, godparents and friends. Their design and manufacture often has to meet stringent standards, and competition in the market demands innovation. The packaging design needs to convey rapidly the intended age group and the workings of the toy, and make a bold statement from the shelf about the product's qualities and the manufacturer's position. Tomy Toys, a recent entrant into the international market (though started in Japan 90 years ago) is a fine example of how these challenges can be met in the context of an international operation.

& Gifts

Similarly, the leisure market offers the packaging designer a wide range of opportunities. It is fast-moving, not only with new products coming to the market but also with rapid changes in how leisure time is spent occurring rapidly. New products such as inline skates (as with Metroblade) carry a new social ethos with them, which the designer has to interpret correctly in order to create a successful design.

Equally, established products in the leisure market need to be refreshed and updated to meet changing demands. The repositioning of Sellotape as a range of products rather than a single brand is a striking example of how well-researched and planned design can achieve the rejuvenation of a household name.

Nicole Fahri Black on Black

GIFTS & LEISURE

Designer Din Design, London, UK
Client Nicole Fahri, London, UK

1 74/75

Type	New design
Product	Female fashion
Material	Printed card
Size	Various
Time	6 months
Elements	Company logo
Market	General female 25–35
Support	In-store displays
Key words	Fashionable

Design Brief

The shopping bag has long been recognized not just as a convenient way of taking loose goods home but also as a form of advertising and promotion. Carrying a bag from a famous design label is a fashion statement in itself. Designing the bag is part of the creation of the whole identity of a store.

The ancillary elements such as catalogues and brochures are an important part of the total statement.

Key Factors

Materials
Specifying quality materials is essential in a design context where fashion is concerned.

P.75

Market
The success of a simple design lies in attention to detail and quality control in manufacturing.

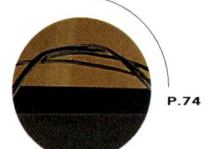

P.74

*' Fashion is today
a brand experience.'*

438mm

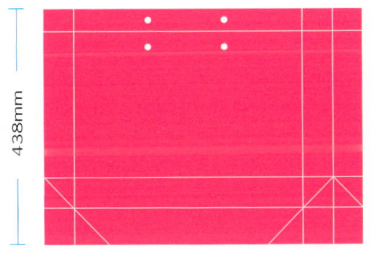

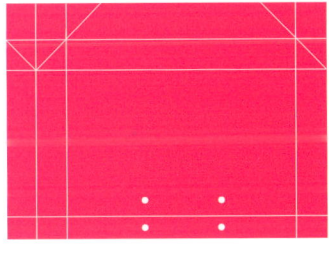

The cutter guide shows how the bag is assembled from two elements for added strength.

NICOLE FARHI

MATERIALS Din Design's work for the fashion designer Nicole Fahri's Black on Black collection ranged from designing labels and price tags, through the menus and paperware for the in-store restaurant and coffee shop, to the bag. The shopping bag is made from high-quality black card, blind stamped and overprinted in gloss back with the collection name in condensed sans serif capitals. The leather thong handles provide a touch of luxury, and the quality of the product is maintained by the addition of a card base liner so that the bag does not lose shape when full. As the cutter guide shows, the bag is made from two main card formers, rather than from a single sheet. This reduces the risk of inaccurate folds and so a poor look to the bag.

The simple bag shape belies its strong two-part structure (see flat plan, above) while the leather thong handles subtly underline the design.

GIFTS & LEISURE

Baby Vision Toys

DESIGNER In-house design by Tomy UK
CLIENT Tomy Toys, worldwide

2 76/77

TYPE	Redesign of existing pack
PRODUCT	Children's toy
MATERIAL	Printed card
SIZE	Various
TIME	Continuous redesign and evaluation
ELEMENTS	Barcode and company logo, safety information
MARKET	General
SUPPORT	Advertising, in-store displays
KEY WORDS	Safe, inviting, educational

DESIGN BRIEF

Tomy Toys is a 90-year old Japanese family business which has expanded worldwide into the toy market in the last ten years. The complete product range is split between toys for the Japanese market only, and toys for both local and international markets. The overall target age range is birth to ten years old, and within the main brand there are a number of sub-brands, such as Baby Vision, and also the very successful Sylvanian Families miniatures range. Current new product developments include children's electronic learning products and software.

'*Tomy offer toys that educate through play.*'

Key Factors

DESIGN
A product which sells equally on look and functionality (like a toy) needs visible display.

P.76

ELEMENTS
In a closely-regulated market information must be clearly accessible.

P.78

MATERIALS
Where packaging is being assembled in different places worldwide, simplicity is important.

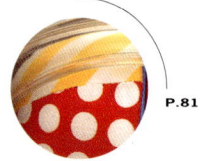

P.81

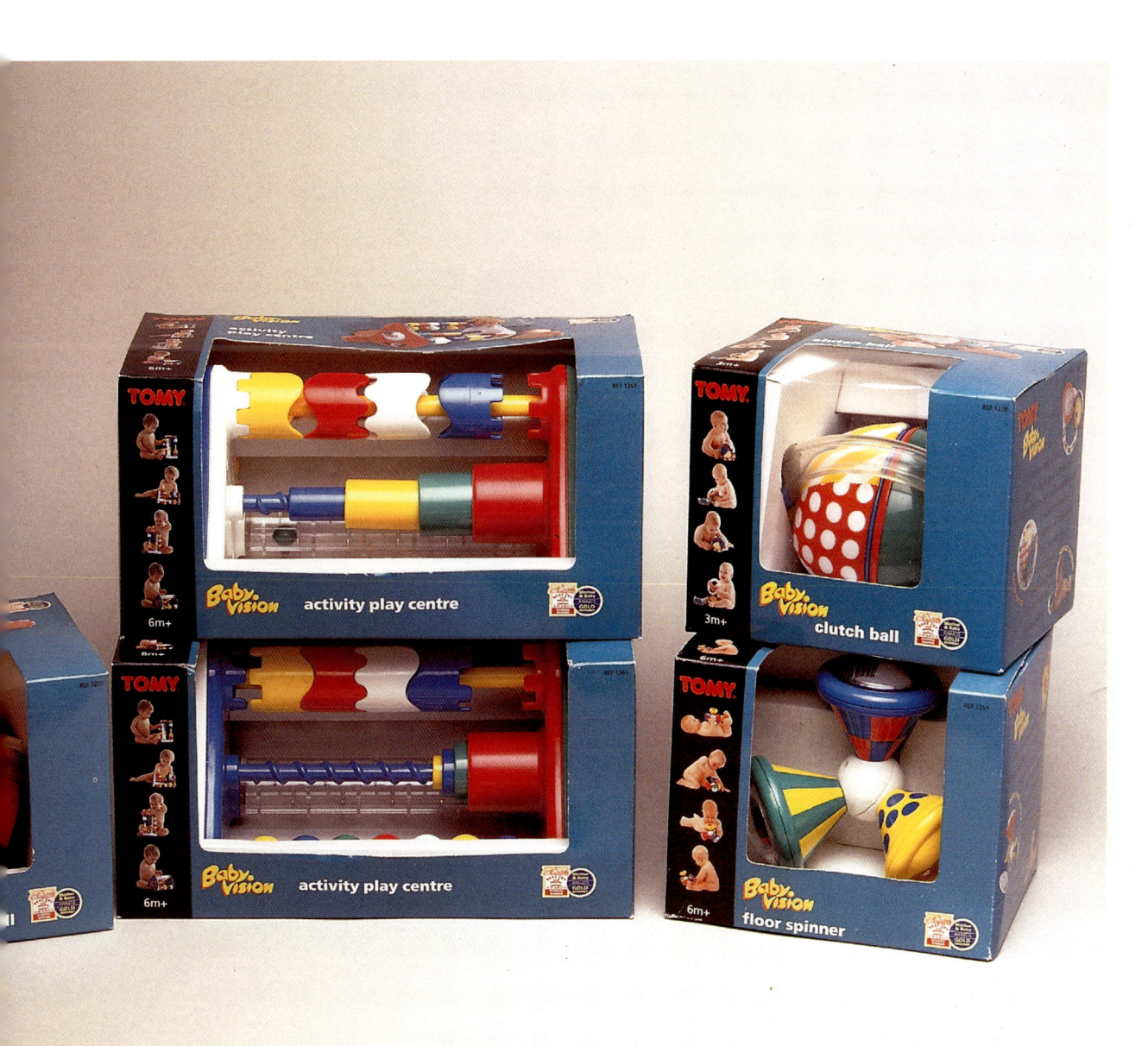

Toys are bought for their looks as much as what they do or teach, so open packaging invites the customer to touch and feel.

GIFTS & LEISURE

Baby Vision Toys – design development

2   78/79

Tomy offer toys that educate through play.

ELEMENTS

Toys for babies and toddlers are bought by parents, friends and relatives rather than by children themselves (this starts with older age-groups). Tomy's success lies in combining visual and fun appeal with a range of functions (movement, sound, tactility, color) which in themselves have educational value. The toys use shapes and forms from traditional toys (balls, wheels, columns) but mix and add effects to them to complete a wholly new range. In effect, the toys are aimed at adult's perceptions of babies' needs and desires. To stress these qualities, primary colors are widely used on the toys themselves, and the packaging is deliberately open so that the customer can try the toys out by touching and shaking them. And because parents are safety conscious (just as are legislators in this field) the statutory information is clearly set out, along with educational values, on the packs themselves.

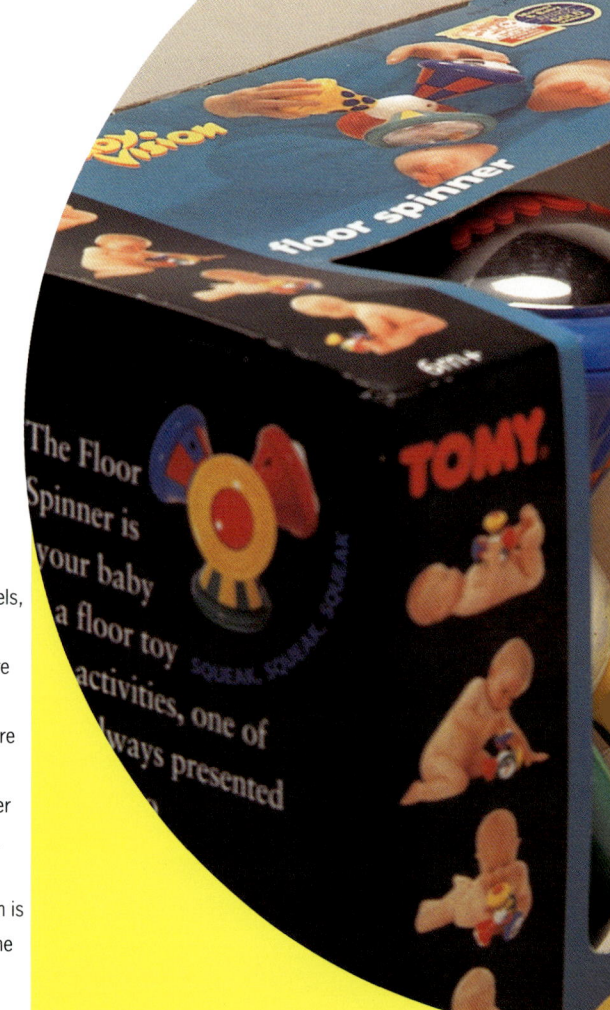

Because they are targeted at a very specific age range, the sub-brand Baby Vision is absolutely appropriate. However, it is always linked positionally on the packaging with the main Tomy name to build brand fidelity as the children get older.

391mm

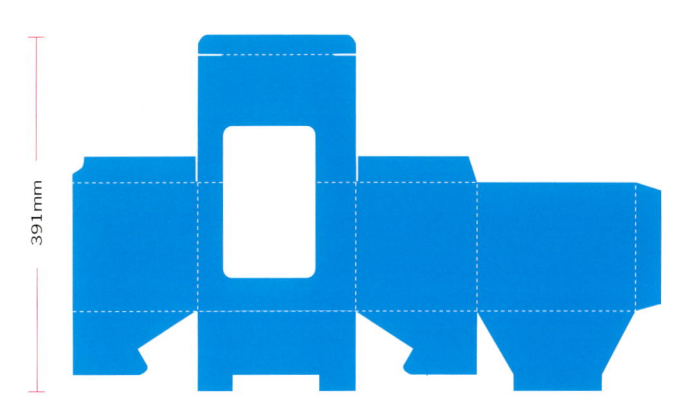

The cutter plan for both box and liner are deliberately simple to allow for different printing and manufacturing locations, and to minimize costs.

The open package approach leaves the frame for presenting key information (main and brand logos, age range, etc.). Only in cases where the product is fabric-covered (such as the Clutch Ball) is a clear protective shell used.

Educational information, necessarily in several languages, is presented on the back, while statutory and recycling information, is on the base.

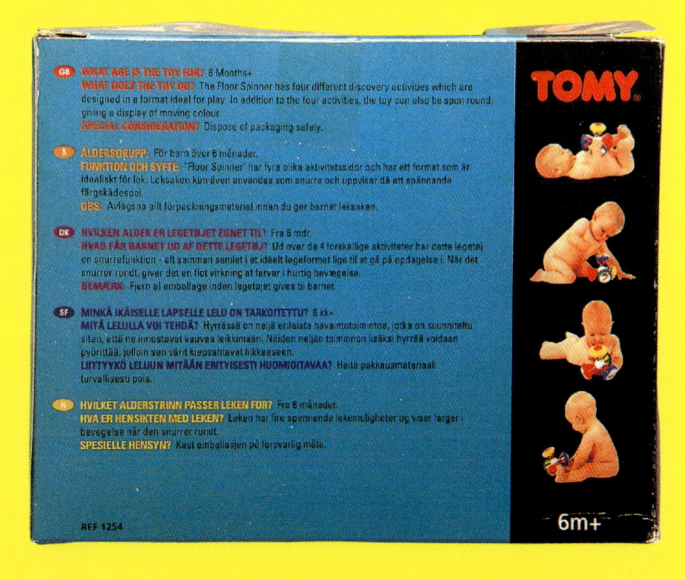

GIFTS & LEISURE

Baby Vision Toys – design development

TYPE	Redesign of existing packs
DESIGNER	In-house team at Tomy UK
CILENT	Tomy Toys, worldwide
PRODUCT	Children's toys
MATERIAL	Printed card
SIZE	Various
TIME	Continuous
MARKET	General

DESIGN DEVELOPMENT

The packaging design is intended to define each range with a key color (turquoise for Baby Vision), but also allows the prospective buyer to examine the product by touching and moving it. The designs work in approximately two-year cycles, according to Chris Jones, European brand development manager for Tomy UK, though this depends also on the market response to new products.

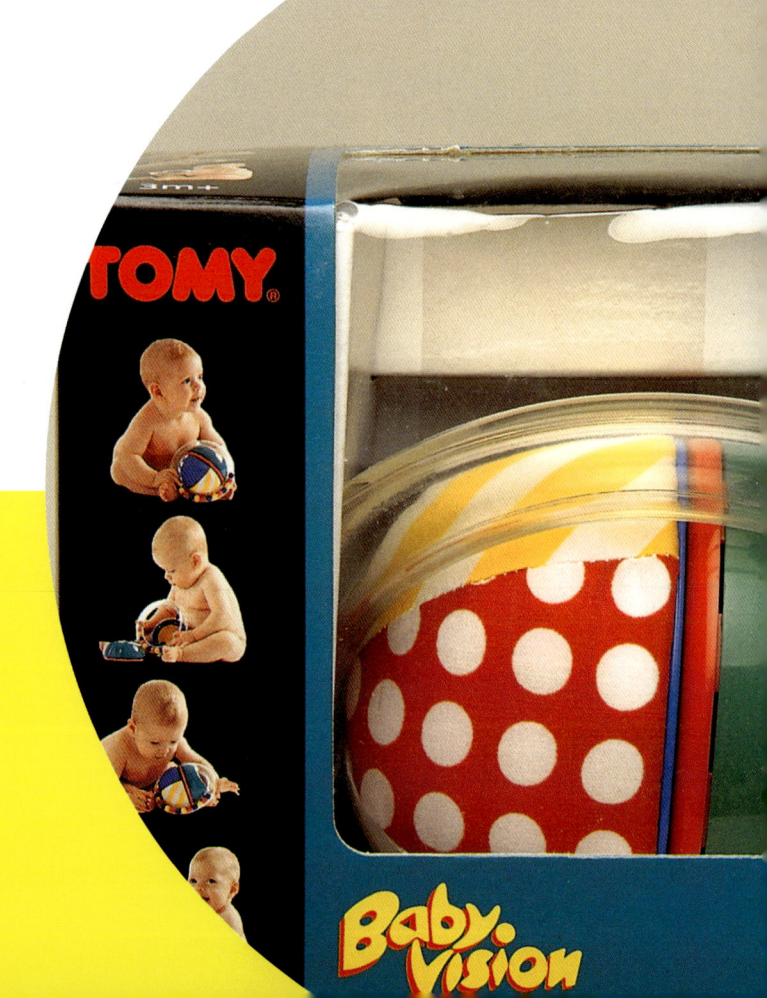

MATERIALS

All packaging design is originated by the in-house team in the UK, but is produced locally wherever the toys are manufactured. This requires a broad general approach, using simple card formers that can be produced anywhere, with clearly visible ties. Since the toys are shipped worldwide in the packaging, it is essential that this be sufficiently robust to stand handling, and it is also designed on a modular system to occupy fixed units of shelf space.

The back panel is dedicated to consumer information, here in a range of languages for the European market.

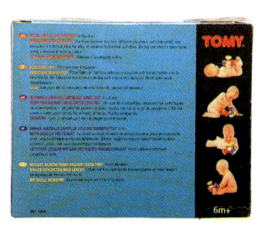

Cellophane is used on the Clutch Ball to protect the cloth surfaces while on display; otherwise open-fronted boxes are the rule.

The pack sizing design **is modular** to occupy **fixed units** of shelf space in twos or threes,

The positioning of the main and subsidiary logos is consistent across the range to build consumer confidence.

The inner liners are of plain white cardboard to show the product to best advantage and to maintain quality control.

Sellotape

Designer The Identica Partnership, London, UK
Client The Sellotape Company, Dunstable, Bedfordshire, UK

' The current strategy provides Sellotape with a long-term platform to develop into new markets and build on existing consumer relationships. '

Type	Redesign of existing brand range
Product	Adhesive tapes for home and office use
Material	Various dispensers including flow-wrap, vista cards, blister packs
Size	Various
Time	12–15 months
Elements	Brand strategy, packaging
Market	General retail, office, home and DIY
Support	Shelf support, mail order, catalogs, trade backing
Key words	Partnership, targeted products, value, quality, innovation

The visual equities of the name, colors and motif were strong but limited to a single sector.

Design Brief

In France a tonic water is called 'un schweppes', in England 'hoovering' is synonymous with carpet cleaning; so one might think that when a product becomes a household name, like Sellotape, the manufacturer has achieved an ideal level of market penetration. However, there can be drawbacks. The company becomes identified with a single brand or product, and cannot easily expand the product range.

Design Sellotape faced two particular problems: firstly they had to re-establish their distinct product offer, as well as increasing consumer awareness of the product range, secondly they needed to find a method of developing into new markets. They approached the design group The Identica Partnership, based in London, to help them find a solution. Identica's analysis of the situation confirmed that whilst the Sellotape brand name commanded a high recall rate, consumers were not differentiating between Sellotape and other competitors' products when purchasing. This situation was compounded by the perception that Sellotape was a routine stationery item.

Key Factors

Design
Moving a brand from established perceptions into new areas and markets is a key function of packaging design.
P.82

Elements
Selecting the correct equities to aid brand development is important.
P.84

Support
Positioning the parent company at a distance can enable the development process.
P.84

Sellotape
Diamond
Ultra Clear Tape
With Refillable Dispenser
19mm x 33m
Sellotape
THE Original EASY TEAR TAPE
With Refillable Dispenser
Sellotape
mm x 15m
Biodegradable Cellulose

GIFTS & LEISURE

Sellotape – design development

3   84/85

DEVELOPMENT

Based on consumer research, Identica created a range of distinct sub-brands, which specifically targeted defined markets. The two key markets identified by designer and client together were the office and the home, including DIY. Identica developed a range of design solutions, using color coding to establish differences and address those specific market sectors. The new sub-brands were defined as Sellotape, which included the DIY range, and Sellotape Office. The Stick It range, with a selection of character-based tape dispensers, was developed to appeal to children (identified as a potential growth area) as a collectable fun item, rather than just a functional product.

In order to allow Sellotape the freedom to develop a broader base, Identica proposed that a new parent company be established, The Sellotape Company. This would provide Sellotape with a strong platform for the development of new, innovative products and separate the portfolio of brands from the somewhat generic perception of the company.

By making the product title (Carpet or Insulating tape) bolder than the Sellotape name, the move to a new area was reinforced.

The Stick It sub-brand's positioning in a separate market justified a special logo.

Designer	The Identica Partnership, London, UK
Client	The Sellotape Company, Dunstable, Bedfordshire, UK
Product	Adhesive tapes for home and office use
Material	Various card based dispensers

Development

The result of the new strategy was to focus the attention of Sellotape's customers on the potential of the brand and to produce a marked increase in sales. The design team not only supplied a solution to the immediate problem, but provided a long-term design strategy for the future. In the words of Neil Ashley, CEO of The Sellotape Company, 'Identica has provided Sellotape with a long term platform for innovation, while enhancing its reputation with existing customers'. Identica's work for Sellotape shows how, in the right hands, strategic brand packaging can change consumer perceptions, and return a significant growth in sales. This was achieved through market analysis, relevant creative solutions and above all through a broad-based approach to the question of a brand's function and its market profile.

The animal figures address the children's market and create a potential equity to be exploited in other ways, such as children's books or TV shows.

THE SELLOTAPE COMPANY

The visually dynamic new logo is a platform for the extension of the brand range.

The current **strategy** provides Sellotape with a long-term **platform** to develop into **new markets** and build on existing consumer relationships.

GIFTS & LEISURE

Orient Tea Room

Designer Alan Chan Design Co, Hong Kong
Client Mandarin Hotel, Hong Kong

4 86/87

'What can be more special to the Orient than both tradition and change.' Tanizaki

Type	Redesign of existing range
Product	Teas and related products
Material	Printed card
Size	Various
Time	7 months
Elements	Colorful logo
Market	General adult
Support	In-store promotions
Key words	Exotic, distinctive, welcoming

Design Brief

The Orient Tea Room offers an exclusive restaurant atmosphere centred around the tradition of a Chinese tea pavilion. They also have tea products for sale to tourists and business visitors to Hong Kong. The packaging had to integrate with the restaurant setting yet be distinctive, as well as matching expectations of Western and Eastern tastes. But the 'fake pagoda and cut-out dragon' approach was felt to be quite wrong. A subtler method, using fine Chinese calligraphy, is used instead.

Support The move to the restaurant is handled by choosing a color for the plates that is as strong as the packaging but different, and linking them through the menu.

The carefully folded boxes subtly suggest the tradition of fine paper and origami, while the overall link is created by a distinctive dark color and the use of gold type.

Key Factors

Materials The choice of packaging material must reflect cultural values, as well as manufacturing requirements.

Support Packaging design can be used to link such areas as service (restaurant) and retailing (fine teas).

P.86

P.86

蘭亭茶叙
LANTIN TEA HOUSE

Metroblade Inline Skates

Designer De Witt Anthony, Northampton, MA 01060, USA
Client Rollerblade, USA

'Travel Yourself. Parking is free.'

Type	Redesign of existing pack plus point of sale
Product	Roller skates
Material	Printed card
Size	Single pair
Time	4 months
Elements	Barcode and company logo
Market	General adult
Support	In-store danglers, press advertising, catalogs
Key words	Streetwise, cool, determined

Design Brief

Just as snowboards have replaced skis, so inline skates have replaced traditional 2x2 roller skates. The change goes beyond the technical into a new cultural dimension: skis are about precision, snowboards about expression, for example. The lifestyles of the products have changed radically. Put more simply, inline rollerskates have moved from being a sports product to a fashion accessory. Therefore they need to be marketed as such, without leaving the sports aspect of physical dexterity behind.

The box-in-a-box approach reinforces the awareness of the product itself by reassuring the purchaser even after buying.

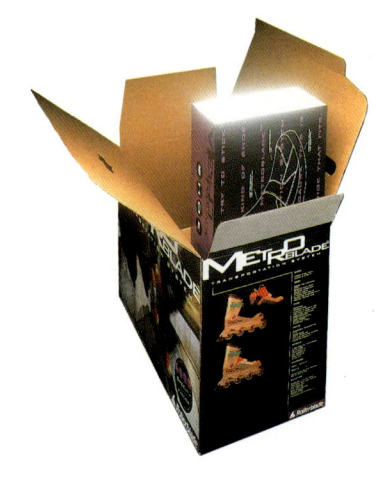

Key Factors

Design
Linking the look of a package to the cultural and lifestyle attitudes of consumers.
P.88

Elements
Choosing graphics and terminology that reinforce end-user aspirations.
P.90

Support
Creating a sub-brand when the main brand name has become a generic.
P.93

The term 'rollerblading' has become a general term for inline skating even though Rollerblade is a company name: the sub-brand Metroblade has been developed to help with this.

Metroblade – design development

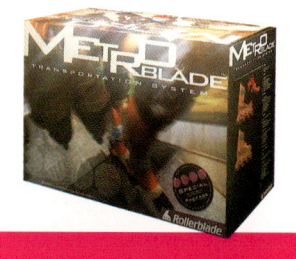

Gifts & Leisure

The strong visual content of the boxes makes them into a permanent display unit in-store.

ELEMENTS

De Witt Anthony's design support for Rollerblade shows a keen awareness of this change. They are responsible for both packaging design and point of sale. That the two go together is not surprising. A fashion item is bought on the basis of its own appearance, while for a sports item performance is equally important. In bringing Rollerblade to the market both considerations had to be judged: the skates were going to be bought both by experienced skaters, for whom performance counted, and by beginners wanting to make their mark with the right accessory. (There is a clear parallel here with sports shoes and trainers, by no means all of which are bought by athletes! Interestingly, De Witt Anthony also number Reebok among their clients.)

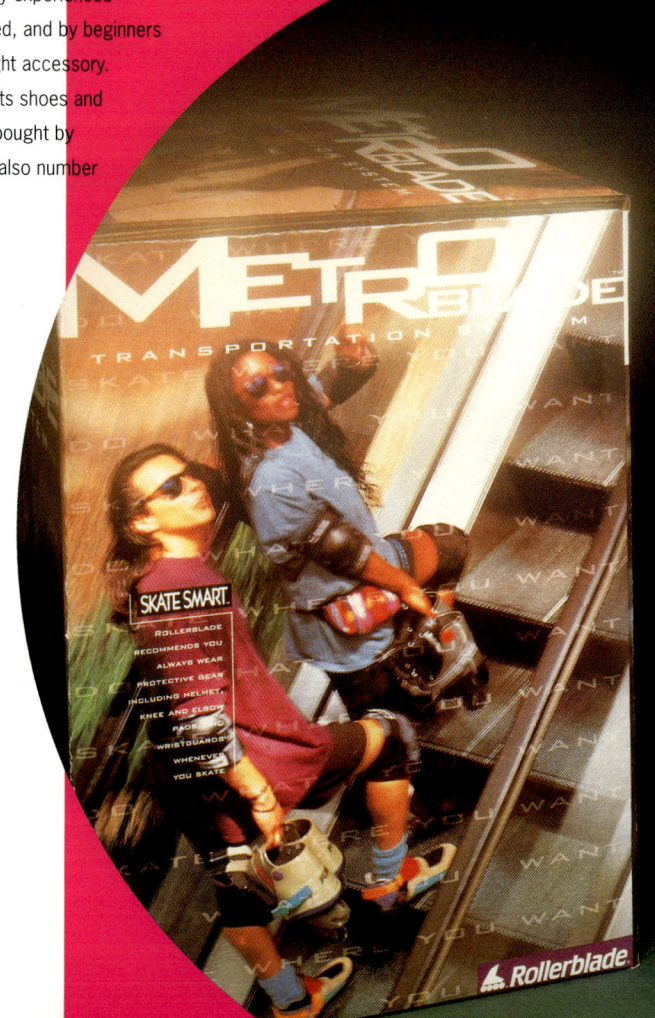

MATERIALS

The normal display approach to fashion is to show clothes directly, unpackaged, in an environment carrying the name and the ethos of the brand. Skates are too large to permit this solution. So De Witt Anthony devised a series of boxes in which the skates would be sold, together with a point of sale display showing the actual product. The boxes are graphically rich, in terms of color and illustration, and overprinted with a series of slogans that both differentiate levels of product and reinforce the fashion aspect. The graphic styling, with exaggerated typefaces, changes of type size and contemporary colors, is perfectly tailored to a market for whom Wired and RayGun is *de rigueur* reading.

travel yourself,
parking is free

Typographic details also echo the dynamic use of lettering on E-zines and web sites, record covers and rock posters.

Action photography overlaid with type brings the look of the pack into line with other media.

GIFTS & LEISURE

The point of sale unit uses an outline design of the back so as to remain functional even if the skate is taken off for examination.

Metroblade – design development

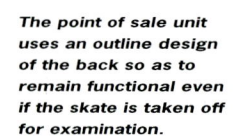

3 | 92/93

MATERIALS

Finally, the point of sale package uses a forward slanting oval as an overall shape to emphasize the forward, positive role of the skates themselves and as a contrast to the rectangularity of the packaging boxes. It is not surprising that this extremely well planned and executed design was honored with a 100 Show award when it appeared in 1994. As a mode for the design student, it shows that a complete integration with the values and aspirations of a product can produce extremely successful results.

Graphic details on the point of sale (or on posters and flyers) do not mimic the packaging but share the same values.

Designer	De Witt Anthony
Client	Rollerblade, USA
Product	Roller Skates
Material	Printed card, acid treated steel

Design Development

The point of sale display unerringly supports the design values of the boxes, and adds another dimension. Firstly, it shows the actual product, mounted in a perforated metal holder (in itself, through the choice of materials, a design statement). This allows the consumer to measure the visual and fashion qualities of the product. The background panel contains the central fashion/sports/independence statement, 'Transport yourself', graphically configured to extract the nuances of meaning in the phrase by emphasizing the word 'self'. The second layer of meaning of the backdrop is found in the white-out texts that highlight the particular technical aspects of the skate: the materials used for the body, the fastening system, the braking bar at the back, and so on. These remain valid even if the skate is removed, through a white outline of its shape and design.

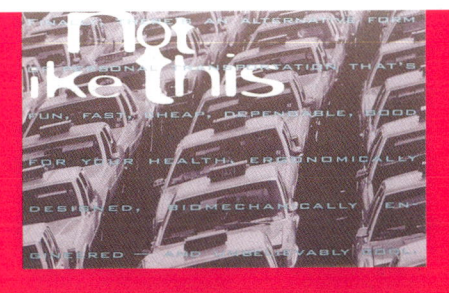

The flyer/poster looks like a spread from Ray Gun, and is unlike the packaging to the point of a new Metroblade logo.

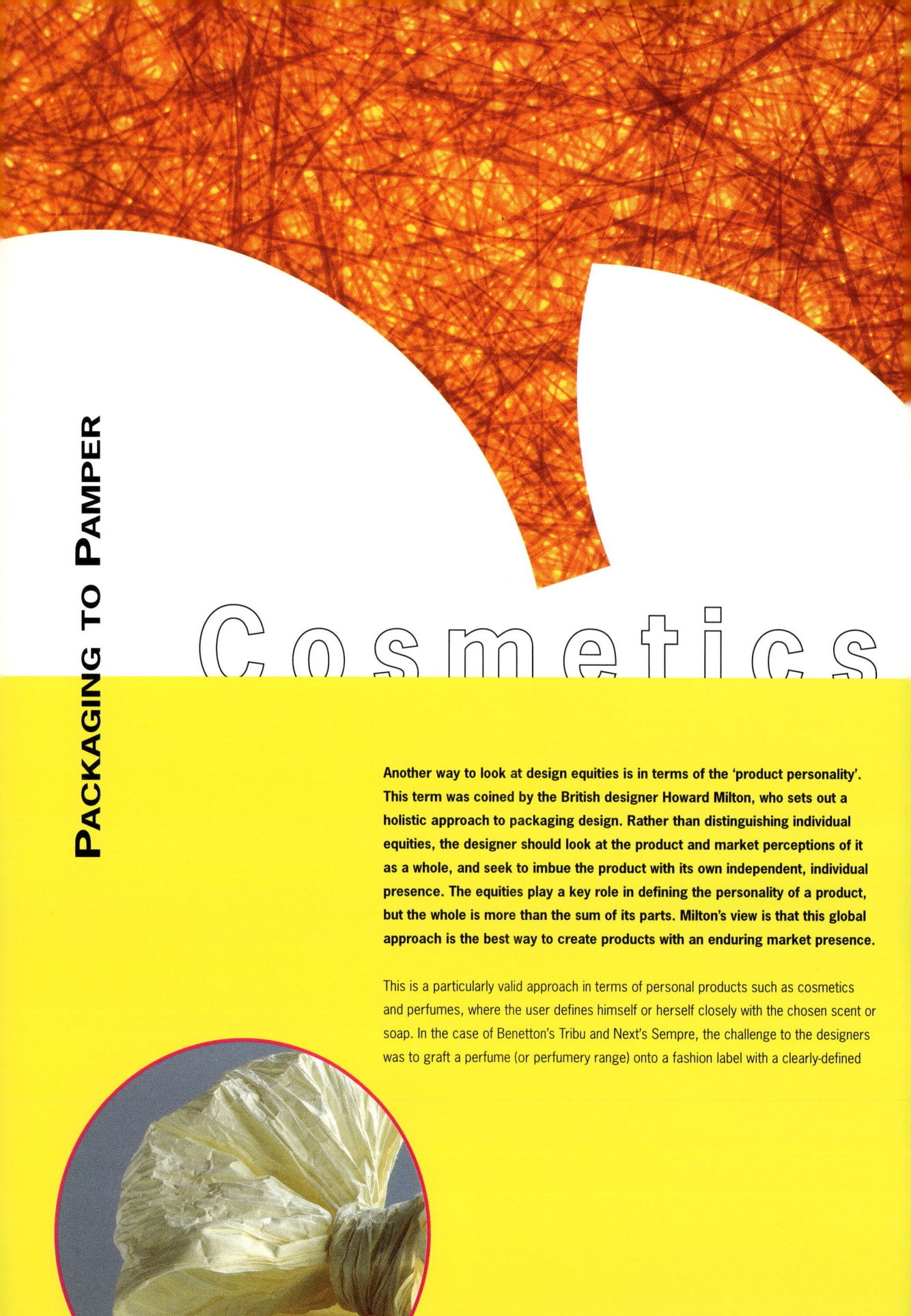

PACKAGING TO PAMPER

Cosmetics

Another way to look at design equities is in terms of the 'product personality'. This term was coined by the British designer Howard Milton, who sets out a holistic approach to packaging design. Rather than distinguishing individual equities, the designer should look at the product and market perceptions of it as a whole, and seek to imbue the product with its own independent, individual presence. The equities play a key role in defining the personality of a product, but the whole is more than the sum of its parts. Milton's view is that this global approach is the best way to create products with an enduring market presence.

This is a particularly valid approach in terms of personal products such as cosmetics and perfumes, where the user defines himself or herself closely with the chosen scent or soap. In the case of Benetton's Tribu and Next's Sempre, the challenge to the designers was to graft a perfume (or perfumery range) onto a fashion label with a clearly-defined

& Beauty

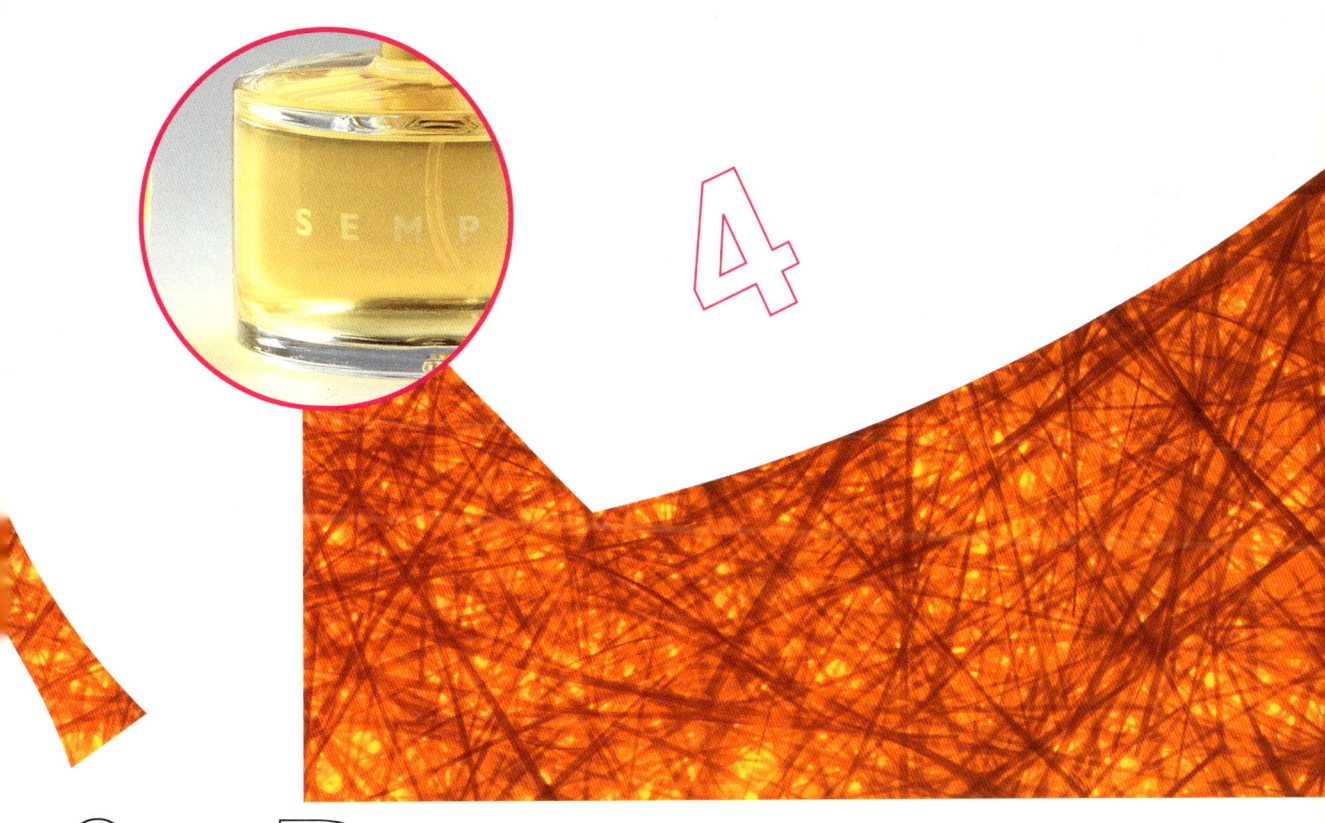

4

personality of its own, with a known target market. In the case of the Body Shop, which had itself created a whole new market for natural cosmetic products using a holistic approach to the body, the problem was refining an extremely successful concept and so keeping ahead of the competition while maintaining the core values of the range.

In all these cases, and with Insignia, the designer's aim has been to convey a social context and attitude through the product, so that it falls naturally into the customer's framework of perceptions.

All packaging should aim to do this, but it becomes a paramount requirement in products to be used on the person of the customer.

Benetton Tribu Gift Set

COSMETICS & BEAUTY

DESIGNER Nicosia Creative Expresso Ltd, New York, USA
CLIENT United Colors of Benetton, Italy

1 — 96/97

The bottle design by Yagi Tomatsu provided a visual starting point for the design.

'*Tribu celebrates the differences in all of us.*' **Flare magazine**

TYPE	Special gift pack for existing perfumes and toiletries
PRODUCT	Toiletries gift set
MATERIAL	Printed recycled card with misted plastic lid
SIZE	14in diameter
TIME	6 months
ELEMENTS	Tribu brand name
MARKET	Female, upper to middle income level, aged 21–39
SUPPORT	In-store promotion, general TV and press advertising
KEY WORDS	Global, environmental, original

DESIGN BRIEF

The launch of a successful packaging design requires many qualities, and chief among them are organization and inspiration. United Colors of Benetton is a worldwide fashion *marque de fabriques*, with headquarters in Italy but retail outlets all round the world. In launching a range of perfumes and cosmetics, Toscani at Benetton sought to endorse the central value of the main brand: international co-operation across races and cultures. Tribu was created as a product range using natural materials from all over the globe, and offering quality at an affordable price for the younger generation. The creation of the brand was wholly international, involving Benetton's head office in Italy, their Paris and New York offices, the product designer Yagi Tomatsu in San Francisco and Nicosia's own creative team (itself drawn from Europe and the Americas).

Key Factors

DESIGN
International designs often require working across national boundaries as part of a broad team. 
P.99

MATERIALS
The choice of materials (especially recyclables) can support the central position of the client, as well as meet the brief. 
P.99

TYPE
Promotional or special items need to balance their special presentation with wider support for and from the main brand.
P.101

UNITED COLORS OF BENETTON
TRIBÙ

The semi-transparent lid provokes curiosity and still hints at the contents.

COSMETICS & BEAUTY

Tribu Gift Set – design development

1 98/99

The gift pack appears to contain three elements, but there is in fact a bonus (below) in the pumice stone placed under the soap.

Tribu celebrates
the differences in all of us.

The materials for the packaging are entirely recyclable, a first requirement of the brief.

Product Type

The creative co-ordination and strategic planning flowed from James Berard at Benetton New York to Tomatsu and Nicosia, following an informal agreement that Tomatsu would have final say on creative work not done by him. 'This was done so we could ensure that the final creative product was pure and consistent,' Davide Nicosia comments. Tomatsu was responsible for the bottle shapes directly, while Nicosia designed the gift set housing. In order to achieve a simultaneous launch across the USA, extremely complex planning and co-ordination with suppliers was required.

274mm

The liner is assembled from two sheets of recycled card cut on the plans shown here.

Materials

The packaging design for Tribu has been acquired for the permanent collection of the San Francisco Museum of Modern Art, and the fragrance products endorsed by PETA, People for the Ethical Treatment of Animals, for their cruelty free development. The totality of product and packaging supports the global approach – 'making the wearer more connected to the world and the people who inhabit it'.

The cutter pattern (above) shows how the recycled card folds to make the inner support liner. The set contains Tribu perfume (in a glass and sanlyn bottle designed by Tomatsu), body lotion, soap and pumice stone (under the soap).

COSMETICS & BEAUTY

Tribu Gift Set – design development

1 100/101

The outer card case, also in recycled and recyclable card, is deliberately understated.

DESIGNER	Nicosia Creative Expresso Ltd, New York, USA
CLIENT	United Colors of Benetton, Italy
PRODUCT	Toiletries gift set

DESIGN DEVELOPMENT

'*The set had to have a unique point of difference: in this case mixing materials to express texture.*'
Maurits Pesch,
Nicosia Creative Expresso

The graphics on the box needed to support the main Tribu graphics, with plain sans serif lettering in white or black. The box shape had to be convenient for in-store display (so the lid fits under the round box). The main box is a section through a redundant cardboard tube (thus saving further costs and use of raw materials) with a folded card former to keep the products in place. The careful paper engineering of this kept material use down to a minimum, as the cutter drawing shows. The outer box is in recycled plain brown card, overprinted in black.

The set had to have a **unique point** of difference in this case **mixing materials** to express texture.

The inner box has a textured look, through raw mix in the card material.

The logo includes the United Colors of Benetton strapline as well as the Tribu mark. This is repeated on the card case and on the translucent lid.

SUPPORT

The design parameters for the Gift Set were: firstly the price range and retail acceptance, secondly the product mix, thirdly the packaging type and finally to support the main Tribu line. Wholly recyclable packaging using SAN plastic and recycled board (in place of the usual plastic liner) was appropriate both in terms of cost and the corporate philosophy. The product mix featured some of the most successful products in the range, but also provided a platform for introducing more special, textural products.

The result achieved a 96% sell through in-store: a considerable achievement given that anything over 35% on a special pack is considered acceptable.

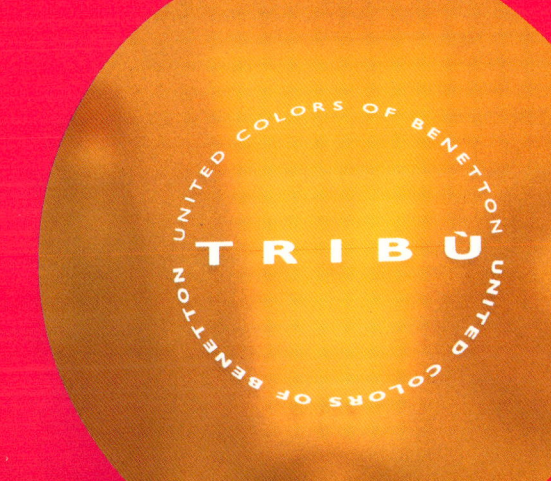

Body Shop Bodycare

Designer In-house design team
Client The Body Shop, UK and worldwide

Type	Redesigns and new products
Product	Bodycare products
Material	Various
Size	Various
Time	Continuous updating
Elements	Barcode, product information and company logo
Market	General
Support	In-store promotions and advertising
Key words	Holistic, environmental, safe

Design Brief

The plain white bottle stressed from the beginning the Body Shop's commitment to a direct, ethical approach: it is now a strong design equity.

The Body Shop is a retailing success story with few parallels. Starting with one shop in southern England committed to making cosmetics and toiletries for the whole body from natural rather than artificial materials, it is now a world wide name with stores in over 20 countries. The Body Shop's new approach to beauty and health forced established companies in the field to follow suit. Developing and strengthening the company's whole image and its brands was partly a question of maintaining a clear policy, partly a matter of adding products consistent with policy, and partly about packaging the products in a way that conveyed the message. This needed constant attention to maintain independence and market share.

'Body Shop created the concept of a total, natural approach to body care. Our task as designers is to maintain, explain and develop that vision.'

The logo has been subtly modified (it is now used in white, for example) but the principal form has been retained, as together with the white bottle, it is a key equity.

Key Factors

Development
An established brand or range can not only extend into new areas — often it is a key way of strengthening the whole.

P.106

Elements
The central equities in an established design need to be identified and reinforced.

P.107

Support
For products in a controlled retail environment close linkage with point of sale is a major bonus.

P.107

The New Skin range uses the same white bottle, here modified by a pump dispenser, and by printing on the bottle rather than labelling.

THE BODY SHOP
tea tree oil
BODY WASH
THE BODY SHOP
tea tree oil
FACIAL WASH
THE BODY SHOP
tea tree oil
BLEMISH STICK
2.5 ml
tea tree oil
BLEMISH STICK
2.5 ml

THE BODY SHOP
tea tree oil
THE BODY SHOP
tea tree oil
THE BODY SHOP
tea tree oil
THE BODY SHOP
tea tree oil
SOAP
100% Vegetable Soap Base
THE BODY SHOP
tea tree oil
SOAP
100% Vegetable Soap Base
THE BODY SHOP
tea tree oil
SOAP
100% Vegetable Soap Base

Body Shop Bodycare – design development

Unlike a supermarket or beauty shop, The Body Shop only sells its own products in its own stores, and its own products are only available in its own stores. Normally a mainstream manufacturer's product is sold in competition with other manufacturers' products in a context controlled by the shop owner. (The development of shops within shops is in part a way of gaining control over the appearance of retail space.) Even a supermarket own-brand is competing on shelf space with others' products. The Body Shop's retailing mix therefore presents particular advantages, but also responsibilities. 'Our special position,' as Jon Turner, head of design at The Body Shop, points out, 'allows us to integrate completely packaging design, point of sale, advertising and the retailing context'. So the packaging design itself can be more directly focused, as the supporting messages about the product's personality are being carried by the shop. Equally, a product that does not fit in perfectly damages the whole.

DEVELOPMENT

The New Skin range (above) maintains equities with the white bottle and black label, but extends it with a pump dispenser in the bottle and screen-printed copy. Overall it could stand easily alongside products from some years ago. 'The Activist range for men (right) is a clearly workable extension of the core brand into a new sector. Maintaining this kind of continuity while extending it is key to the continuing success of The Body Shop over the competition which has brought plenty of look-alike products to the market.'

The new Tea Tree Oil range uses both the standard bottle and new shapes: the natural source of the oil features in the background for point of sale support.

THE BODY SHOP

Elements

Part of the continuous reassessment of design includes key elements such as the logo, which has been subtly redesigned to give it greater visibility and consistency, and for print efficiencies. The color range of the packaging has also been extended, and individual market segments addressed. The other main design equity – a plain, pale, translucent and recyclable plastic bottle, has been maintained, and additional shapes introduced.

The new Green Label range uses a clear cellophane wrap-around label with the logo reversed out in white, but keeps to the well-established translucent bottle.

Orchid & Calendula WATER
NORMAL TO DRY SKIN
Freshens, soothes and moisturises

The Body Shop created the concept of a total, natural approach to body care. Our task as designers is to **maintain, explain** and develop that **vision**.

Support

From a position of strength, of course, it is possible to advance. The Activist and No Debate products for men (left) share the overall company ethic in product content and development (natural products, and no animal testing, for example), but their marketing approach and naming is more in line with a traditional men's toiletries range. The Body Shop logo is now a sufficient guarantee of the products' honesty and purpose.

Insignia

Designer Blueberry Design, London, UK
Client Procter and Gamble, Weybridge

Type	Redesign of existing pack plus shipper pack
Product	Man's toiletries
Material	Printed and embossed card
Size	2 x 175ml bottles
Time	4 months
Elements	Barcode and company logo
Market	Males 16–29 years old
Support	Seasonal TV and press advertising
Key words	Powerful, controlled, male

Design Brief

Male toiletries offer alternative paths. One is to find parallels to traditional female perfumes (Calvin Klein's recent CK unisex perfume is perhaps the logical conclusion to this). The other is to establish a male vocabulary for the product, as was the case here. The brief at the time called for a typology the target market of young males would identify with. The previous packaging, designed some three years before, has used sport as a metaphor. Blueberry's Russell Sellers felt this was dated, and a more contemporary image was needed. The recent success of film's such as Robocop and the Terminator series suggested a new departure. Since the design of the bottles themselves could not be changed, the colorways of the bottles were used to frame a silver and blue card packaging using embossed mechanical imagery, which deliberately recalled the bionic elements of the hero figures in the films.

Support This approach was reinforced by the advertising support for the product. While the design successfully seized the time – something that was reflected in the sales – it was also liable to date as new fashions of male identity came to the fore. Since the structural packaging of the bottles was not involved in the redesign, only the cardboard outer, this was a perfectly acceptable consequence. In fact, the rates of redesign in such a fast-moving product sector are often faster than in the main retail sector. Here the Christmas gift market plays an important role. It represents a considerable sales opportunity, and manufacturers are happy to invest in topical or immediately relevant packaging to meet demand during this key season. Blueberry's acceptance of this tightly-timed challenge is a good example of how packaging design needs, on occasion, to meet particular opportunities.

Key Factors

Design
Identifying the correct metaphor for the market can create success. 
P.108

Materials
In a fast-moving market, redesigning casings regularly may be a better option than structural redesign. 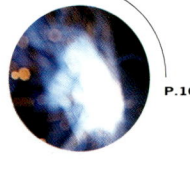
P.108

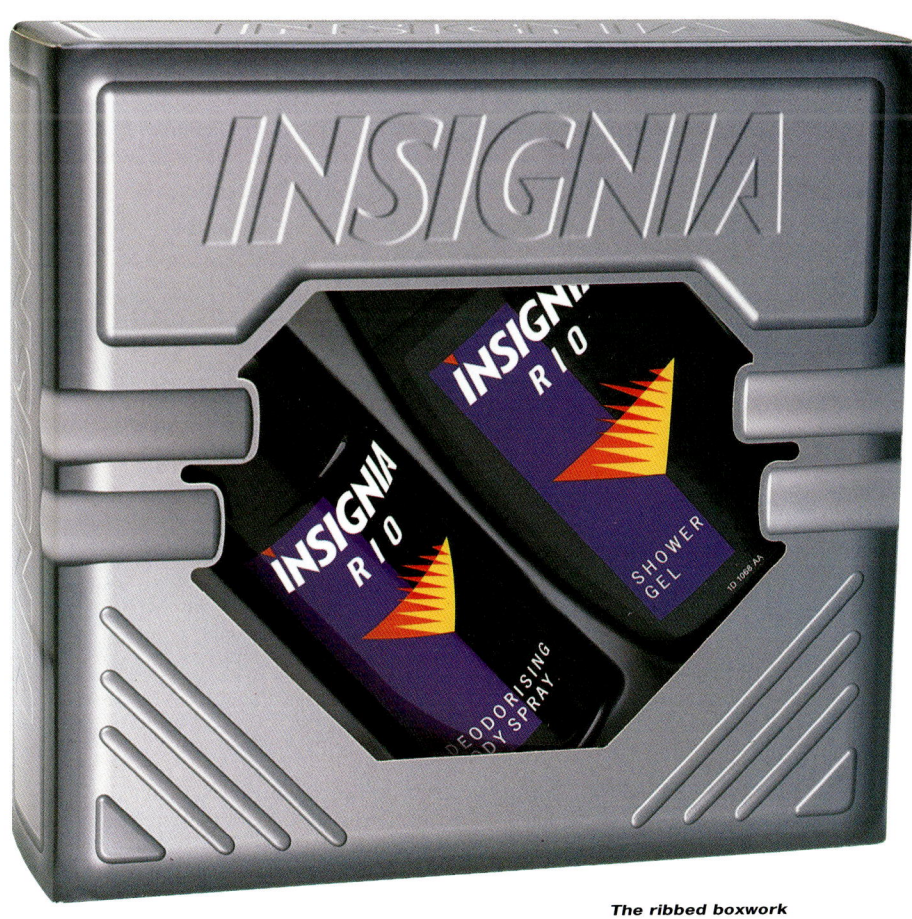

The ribbed boxwork deliberately echoes the prosthetic cybermen of **Terminator** *or* **Robocop**, *a metaphor reinforced by the advertising (opposite).*

Sempre

Designer Lewis Moberly, London, UK
Client Next, London, UK

TYPE	New product range
PRODUCT	Perfume
MATERIAL	Printed card and paper ribbon
SIZE	5ml bottle
TIME	7 weeks
ELEMENTS	Company logo (on base of pack)
MARKET	Women aged 24–30
SUPPORT	In-store promotions, catalog
KEY WORDS	Original, discreet, elegant

Design Brief

The UK fashion chain Next, successfully established in the 1980s by George Davies, created an immediate and considerable following among young career women, aged 24 to 30, and later for men. The company started by mail order sales, modelling their approach not on the down-market value for money level of some UK mail order houses, but rather on more up-market French, German and American mail order businesses such as Les Trois Suisses. The Next catalog itself rapidly became a fashion item! Next also created a range of shops selling their clothes and accessories. Since they wanted to offer their clients a complete fashion range, the decision was taken to add a perfume to the Next Collection, and the packaging design and naming was entrusted to London design team Lewis Moberly.

' The new consumer is more discerning and will not be overawed by image. I take the attitude that retail is part of the communication business. '
George Davies, founder of Next.

The Sempre box itself is carefully understated, and relies on the white bow to make an impact.

Key Factors

MARKET

A single product in a competitive market is often helped by directness and simplicity.

P.112

MATERIALS

Using standard elements such as bottles and boxes is often necessary if design time is short.

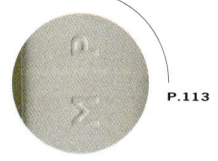

P.113

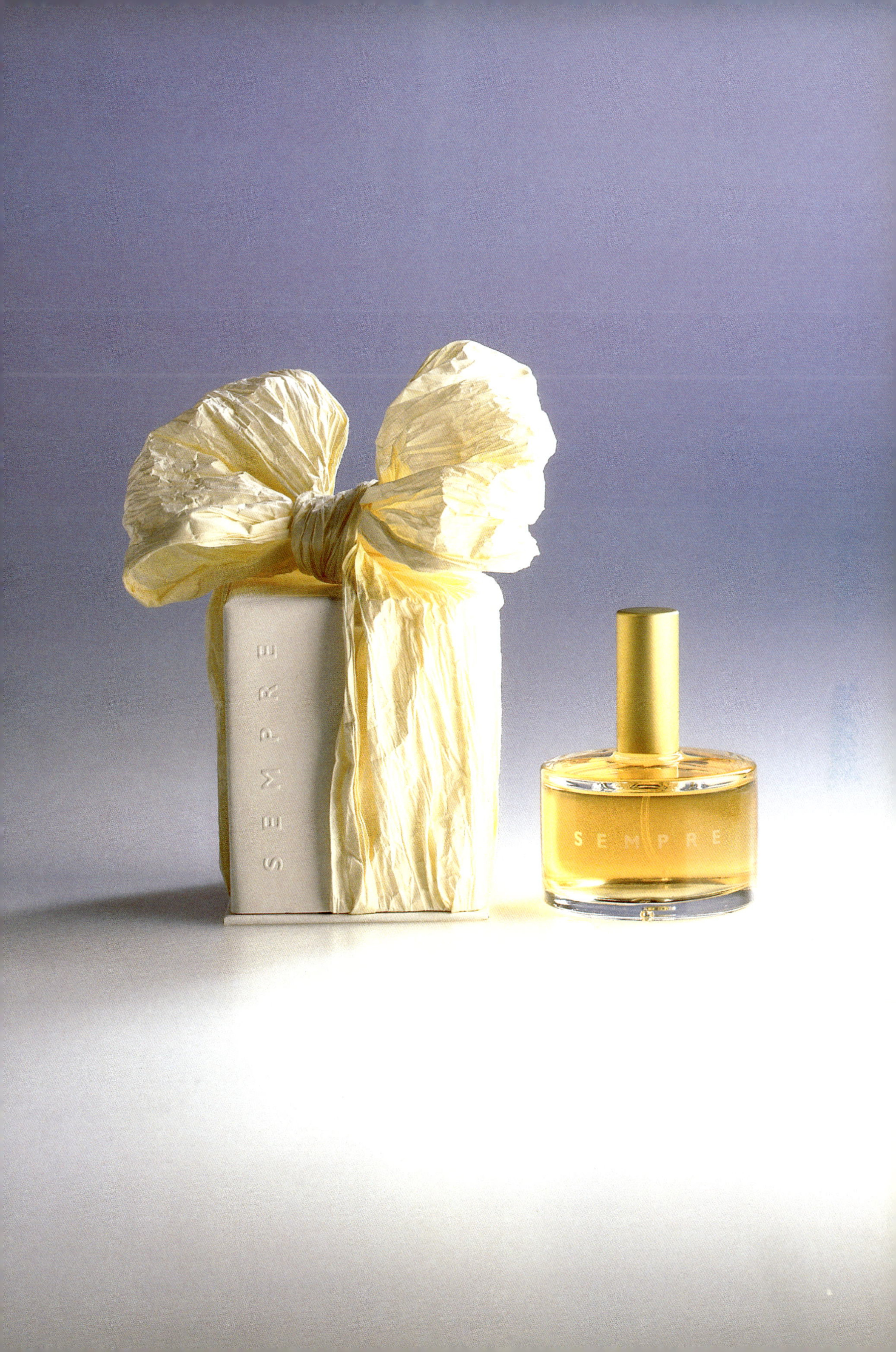
SEMPRE
SEMPRE

MATERIALS

With the cosmetic and perfumery market dominated by established international names, a new perfume has to fight hard – and expensively – for visibility and shelf-space. It also needs visual independence – a new perfume that looks too much like a competing product is dismissed as a clone in short order. As Next had their own catalog and shops, shelf-space competition was not a problem. But their customers were visually highly literate, and aware of the choices among competing brands. Their first perfume had a formal look that was almost drab: Lewis Moberly felt they had to make a stronger and more individual statement, though within the cool expectations of the customer. As the time-scale was very tight, they adopted an existing bottle shape. The outer packaging was an immaculate white card box with rounded corners, slotting onto a foot that held the bottle. The name, chosen by Lewis Moberly, was blind stamped vertically up the side of the box. (Sempre means 'always' in Italian.) Both name and packaging had the right, measured tempo for the market. For the individual gesture, a white paper bow wraps and flamboyantly crowns the box. But the bow is made from unravelled Chinese paper string: it absorbs light, rather than garishly reflecting it, like the ribbons on a floral bouquet or chocolate box. Even here, overstatement is restrained.

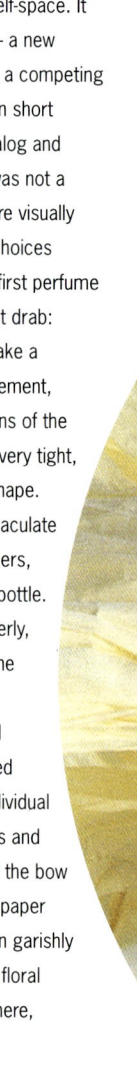

MARKET

The combination of these elements – the understated name on a plain, but meticulously finished, box and the grandiloquent bow made of a wholly basic material – is what makes the Sempre design so successful. The balance of judgment shown by the designers in handling the brief was proved by the success of the new perfume on its launch.

A sans serif face is used for the product name on both the bottle and the box. On the bottle it is condensed.

The bottle is a standard product, used because of the time frame, but saved by the carefully letter-spaced typography.

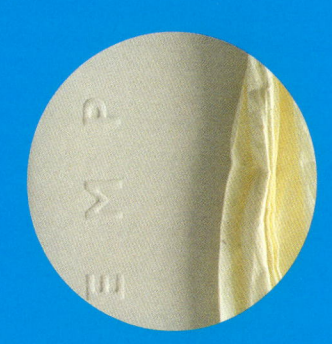

A full width version is used on the box, since the relief blind stamping would not be sufficiently clear in a condensed face. (In blind stamping the type is stamped into the surface without ink; intaglio stamping presses the letters into the surface, relief raises them (by stamping the reverse of the support).

I take the attitude that **retail** is part of the **communication business.**

George Davis, founder of Next.

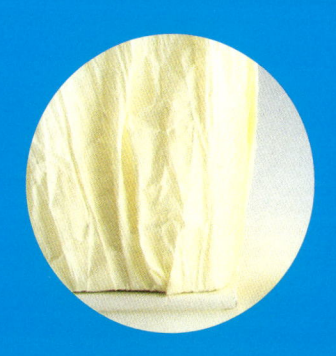

The bow (see main picture) is made from unravelled Chinese paper string (normally found as handles on paper bags). It has the right matt quality to offset the understatement of the whole design.

SELLING TO THE CONSUMER

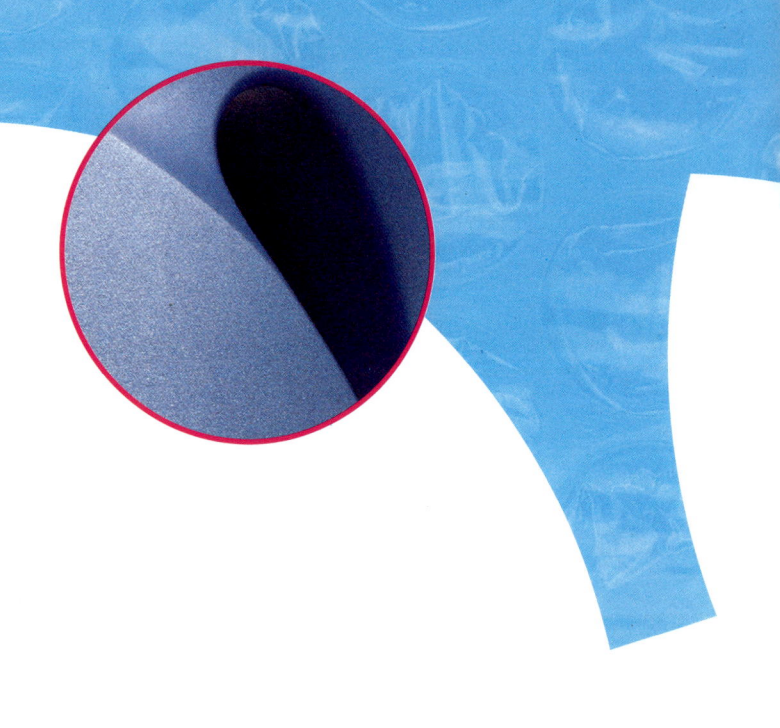

Health

Often the packaging base material is decided by the manufacturing process of the product, as well as by retailing convention. But there are opportunities at times to exploit the packaging material to make a statement about the function of a product, and help define the product in the public eye.

Toilet Duck, although designed some years ago, is a case in point: its quirky shape fits with its name and function into a complete whole. It would be difficult to imagine what could be added to the product – or subtracted from it – that would improve or strengthen its personality.

The use of specially-designed bottles has increased recently, since European legislation, and an American court case over the Jif lemon-shaped lemon juice bottle, has allowed

5

& Houseware

bottle shapes to be copyrighted, along with other visualqualities, such as specific corporate colors. This gives a potential design equity legal protection against lookalike competition: the previous legal remedy was 'passing off', claiming that a competitor deliberately imitated the appearance of a product to deceive purchasers. This proposition was notoriously difficult to prove.

Loewy's work for Simoniz takes full advantage of the potential of the new legislation, while also showing how well-researched design can pay for itself through economies of scale in manufacturing. Oropharma, the third case study here, decided that for a niche market – the carrier and racing pigeon fraternity – standard packs were acceptable, but used a series of intelligent graphic devices to locate each product in the overall range.

HEALTH & HOUSEWARE

Toilet Duck

DESIGNER In-house team
CLIENT Johnson & Johnson worldwide

1 116/117

TYPE	New product
PRODUCT	Toilet cleaner
MATERIAL	Flexible plastic bottle
SIZE	500ml pack
TIME	4 months
ELEMENTS	Barcode and company logo
MARKET	General
SUPPORT	Advertising
KEY WORDS	Safe, hygienic, easy to use

DESIGN BRIEF

The story of the development of Toilet Duck is one of the classics of packaging design. Johnson & Johnson, the American pharmaceuticals group, had developed a toilet cleaner. The question they put to their designers was how to package it so that the user could squirt the product under the rim of the toilet bowl, where it would do the most good. The first option, a flexible carton with a straight nozzle, meant the users putting their hands into the bowl, with the risk of being splashed by the product. A delivery system that obviated this was desirable.

MATERIALS The solution was the reverse angled spout, joined to the main bottle by a neck, to help control the flow. Using this arrangement, the package could be held above the bowl, with the spout pointing under the rim. Squeeze the upper end, and the cleaning liquid was delivered to the right place.

This design solved the delivery problem: it also gave the product its name, as the package, in profile, looked just like a duck's neck and head. The new packaging cost no more to manufacture than a standard plastic bottle, and, because of its unusual but practical shape, was an immediate success.

Key Factors

DESIGN

A special bottle-shape, related to function, can generate a whole new product.

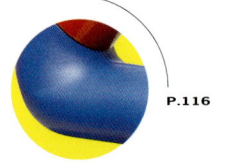

P.116

Once one knows what the product is for, its shape explains immediately the method of use and creates the metaphor for the name.

Oropharma

DESIGNER Kan Design Consultants, Antwerp, Belgium
CLIENT Oropharma NV, Sint-Niklaas, Belgium

TYPE	Design of packs plus branding
PRODUCT	Medicines for pigeons
MATERIAL	Printed plastic containers and card boxes
SIZE	Various
TIME	Continuous project
ELEMENTS	Barcode and company logo
MARKET	Veterinarians and pigeon fanciers
SUPPORT	Specialist magazines
KEY WORDS	Safe, reliable, scientific

Design Brief

Oropharma produce a range of medicines and vitamin supplements for racing and carrier pigeons. These are purchased in part by specialized vets and in part by owners. Kan have been involved in the redesign, repositioning and development of the brand for the last five years. 'It is a continuous process,' according to Hans Kan, partner in Kan Design Consultants in Antwerp.

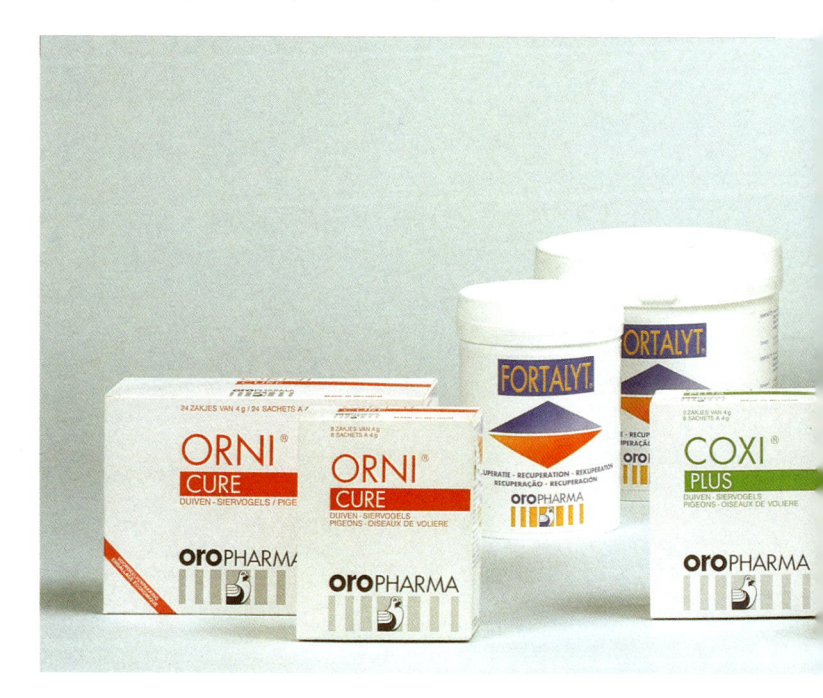

Key Factors

DESIGN

A product for a very specific market can still benefit from good packaging.

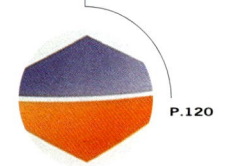

P.120

DEVELOPMENT

A simple visual language of shapes helps define the function of products in a brand.

P.121

The logo highlights the etymology of the brand name, while the birdcode/barcode acts as a visual metaphor.

oroPHARMA

Consistent placing of typographical elements allows the range of medicines to be read as a whole.

The designs are based on standard plastic containers with printed labels. A series of differently-shaped and shaded color patches identify the different groups of products. 'Our intital contribution,' according to Hans Kan, 'involved the design of a line of packaging which had to be innovative without scaring off the rather conservative target group, homing-pigeon *aficionados*. In addition to these packaging assignments, KAN also developed a series of supporting brochures, direct mail campaigns and advertisements – all in the same style. 'Oropharma was able to position itself in the market-place in part due to our design.'

Oropharma was able to position itself in the market-place in part due to our design.

Design

KAN's solution was to use white packaging boxes and plastic bottles, which conformed to industry standards for pharmaceutical and veterinary packaging. They also established a set of design rules. The composite logo – the word oropharma set in a mixture of caps and lower case, accompanied by a series of boxes featuring a pigeon graphic – always appears on the baseline. Medicines have a purely typographic look. The name, set in sans serif caps, surmounts a colored bar which bleeds off to the right. These medicines are color-coded on the bar for the part of the pigeon's anatomy they treat.

Point of sale material shows the product range and its application to the bird.

The different visual clues (triangles, trapezoids, sections) build into a complete family of remedies, each related to a different part of the bird's physiology.

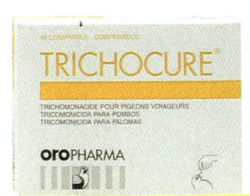

MATERIALS

On the vitamins and conditioners, the same typography, but with the name reversed out of the bar, is also used. A similar color-coding is accompanied by a graphic element, a triangle, arc or double arc printed in the key color, and fading to white at one extremity. This patterning suggests the color patterns on a bird's spread wings, and so acts as a further visual clue to the purpose of the product.

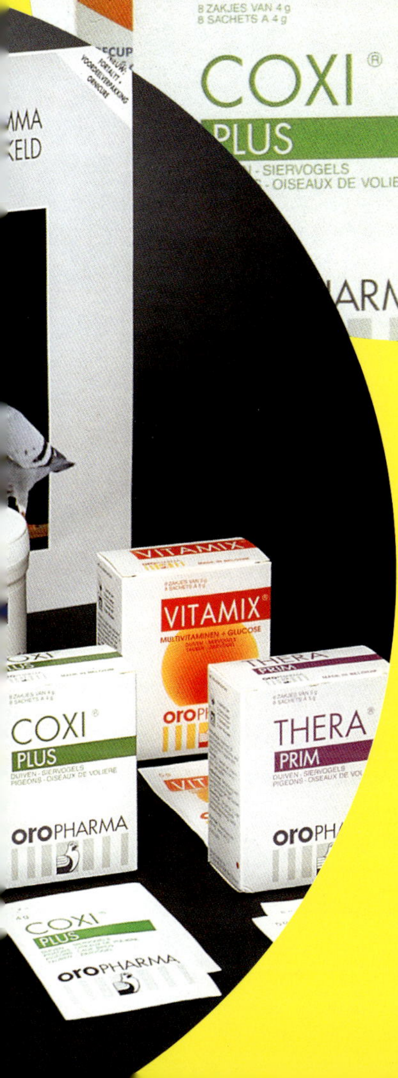

An analysis of the use of margins, rules, cap letters and logo-placement shows how consistently the design has been applied.

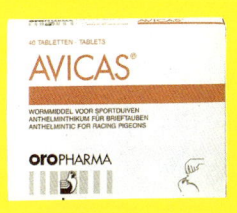

Simoniz

Designer Raymond Loewy International, London, UK
Client Burmah Castrol plc, UK

122/123

Type	Redesign of product range
Product	Car cleaning and maintenance products
Material	Polythene bottles
Size	Various sizes
Time	7 months
Elements	Barcode and company logo
Market	General motorists
Support	General advertising
Key words	Complete, practical, professional

Design Brief

The concept of a design equity has a number of meanings. It can refer to the consumer perception of a design package accrued over time – a kind of visual goodwill. (The fact that this cannot be quantified exactly does not mean that it is not important.) Or it can mean a specific visual quality that can be protected legally, like a visual copyright, as being the property of a particular manufacturer. With the introduction of legislation in Europe in the early 1990s allowing the copyrighting of shapes and designs, part of the challenge for packaging designers has been, where appropriate, to create new forms that could be protected, so preventing competition from look-alike designs.

The previous range of designs used a confusing collection of over 20 different shapes.

Key Factors

Materials
Unifying the packaging base in material-terms can create major cost savings.

P.124

Type
The redesign takes advantage of new design equity potentials.

P.125

The new designs work as a coherent family of shapes.

Health & Household

Simoniz – design development

3   124/125

The new shape is specific to Simoniz, protected by legislation and also identifiable on the shelf.

MATERIALS

At the same time as the new bottles were designed, the main logo was simplified and visually strengthened, as well as being integrated into the bottle shape, again linking the product shape with the brand. Cost savings of 35 per cent are anticipated from the introduction of the new range late in 1996.

The shape is ergonomically planned to fit well into the user's hand.

A selection of sizes and applicators allows the design to cover the product range.

Simoniz were looking to **order** their design range, and gain **equity** in a registered **shape.**

The logo is physically incorporated into the design during the moulding process.

TYPE

When Simoniz approached Loewy International about the redesign of their product range, the idea of creating a specific Simoniz shape was very appealing. Until then Simoniz had been using over 20 different shapes, with consequently high production costs. Loewy proposed as an alternative a range of five linked shapes, which could meet all the requirements of the range and link the products under a single brand.

6

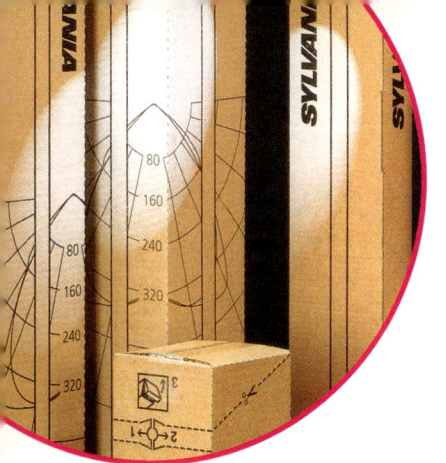

Selling the Hi Tech Way

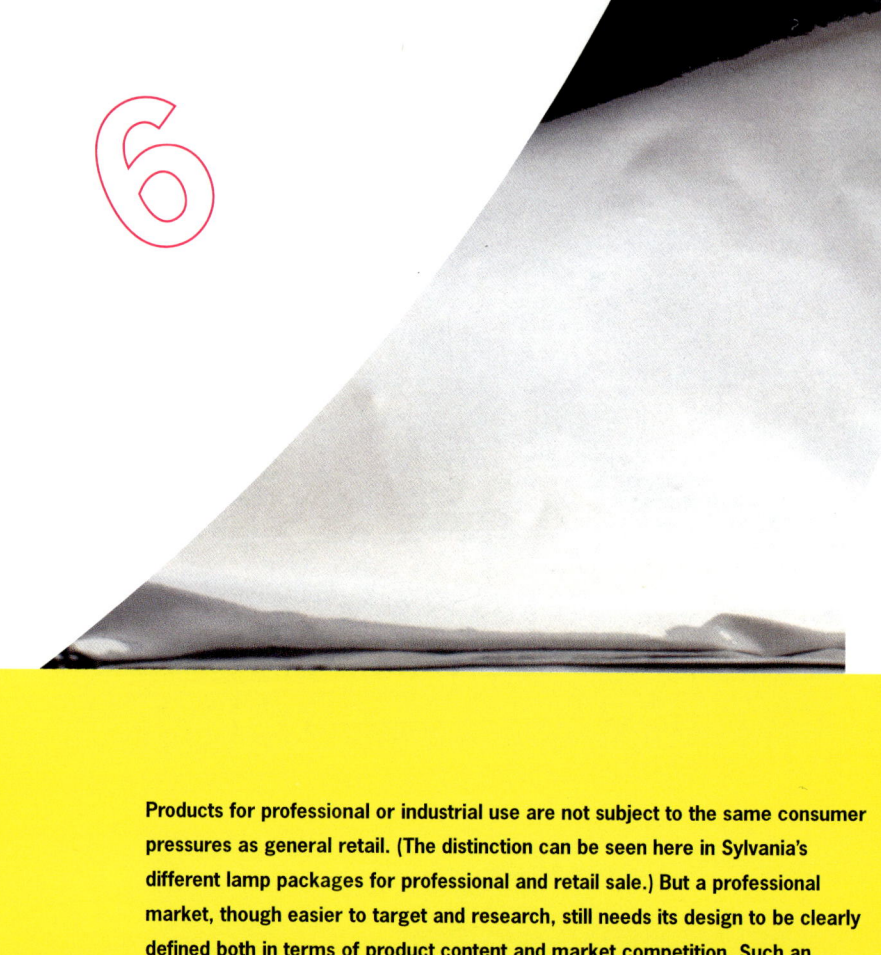

Products for professional or industrial use are not subject to the same consumer pressures as general retail. (The distinction can be seen here in Sylvania's different lamp packages for professional and retail sale.) But a professional market, though easier to target and research, still needs its design to be clearly defined both in terms of product content and market competition. Such an understanding of the approach of the potential professional consumer can be seen with Fractal Design's Painter software package. They quickly realized that their target market – graphic designers and illustrators – had the wit to understand the joke, and also were reassured by seeing new technology in a familiar context. And good design can provide added commercial value: Sylvania's lamp boxes use recycled card, so making a wider point about the environmental policies of the company.

Technology

Such a close understanding of the market can allow an individual design to break ranks from a product range, as with Olympic Weathering Stain, which was seen as having both a professional and an individual market, though a niche in both cases. The added cost of producing an independent design is recouped through broadening the market to new users. Finally, a new metal and plastic packaging system from Germany, Keggy, serves as a reminder that engineering innovation can lead to design and market innovation as well.

TECHNOLOGY

Painter Software

DESIGNER Jenkins & Hal Rucker, Rucker Design Group, Mountain View, CA, USA

CLIENT Fractal Design Co, Aptos, CA, USA

1 128/129

' Painter's array of artist's tools work just like their real life counterparts. '

TYPE	New design (now in fourth revision for upgrades, etc)
PRODUCT	Computer software
MATERIAL	Metal tin with printed paper label, embossed lid and handle
SIZE	5 litre tin
TIME	4 months
ELEMENTS	Barcode and company logo
MARKET	Professional designers and graphic artists
SUPPORT	Specialist advertising
KEY WORDS	Witty, creative, new

DESIGN BRIEF

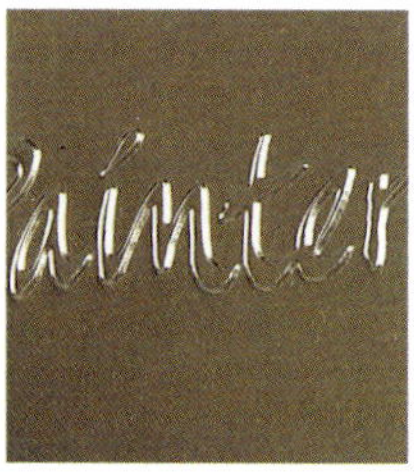

Computer software is largely anonymous as a product: the actual element is a bunch of discs, or, increasingly, a CD-ROM. The packaging therefore plays a key part in helping the consumer identify the qualities involved in the product. By moving away from the traditional rectangular card box used by so many software manufacturers, Fractal Design sought to create a different identity for their Painter software. The choice of a design based on a tin of paint is a neat visual metaphor which suggests that using the software is as simple as applying paint.

Less the paint tin solution seems too evident, the product name has also been reverse-stamped onto the lid.

The wit is reinforced by using a digitized image of a paint tin as the central motif on the label and on the manuals. These are spiral bound for ease of use, and packed inside the tin.

Key Factors

DESIGN

A witty package can sometimes convey the central message of a design very easily.

P.128

Painter's array of artist's tools work just like their real life counterparts.

Computer software is an intangible product: the double metaphor of paint tin and image of paint tin makes this intangibility visible.

TECHNOLOGY

Sylvania Lighting

DESIGNER Premsela and Volk, Amsterdam, The Netherlands

CLIENT Sylvania Lighting International, Geneva, Switzerland

2   130/131

TYPE	Redesign of existing pack range across several companies
PRODUCT	Electric lamps
MATERIAL	Printed recycled card
SIZE	Various
TIME	4 months
ELEMENTS	Barcode, product label and company logo
MARKET	Professional lighting installers
SUPPORT	Catalogs and brochures
KEY WORDS	Professional, user-friendly, part of a total solution

DESIGN BRIEF

Sylvania Lighting International produces lamps and light fixtures for the European, Australian and Middle and Far Eastern markets. It has grown by acquiring a number of existing lighting companies, notably Lumiance in the Netherlands and Concord Lighting in the UK. Its products are therefore known under different brand names in different parts of its total market, even if the products are the same. Part of their sales are in the retail trade, where the lamps have a conventional printed box packaging using the corporate colors of green and gray. But the largest part of their lamp sales is directly to electrical installers, through wholesalers. The packaging for these does not need to have the same shelf appeal. It has to be quickly identifiable by the wholesaler making up an order, and the technical information the user needs must be easily accessible.

' Our packaging design forms an integral part of our overall corporate strategy, to make Sylvania the one-stop shop for lighting solutions. '

Key Factors

DESIGN
Understatement can be successful in a wholesale context. 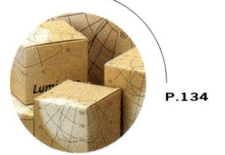
P.134

ELEMENTS
A new logo, as here, needs to be carried through the whole product, packaging and support range. 
P.134

DEVELOPMENT
Retail markets may require a more positive design stance, without losing the established elements. 
P.135

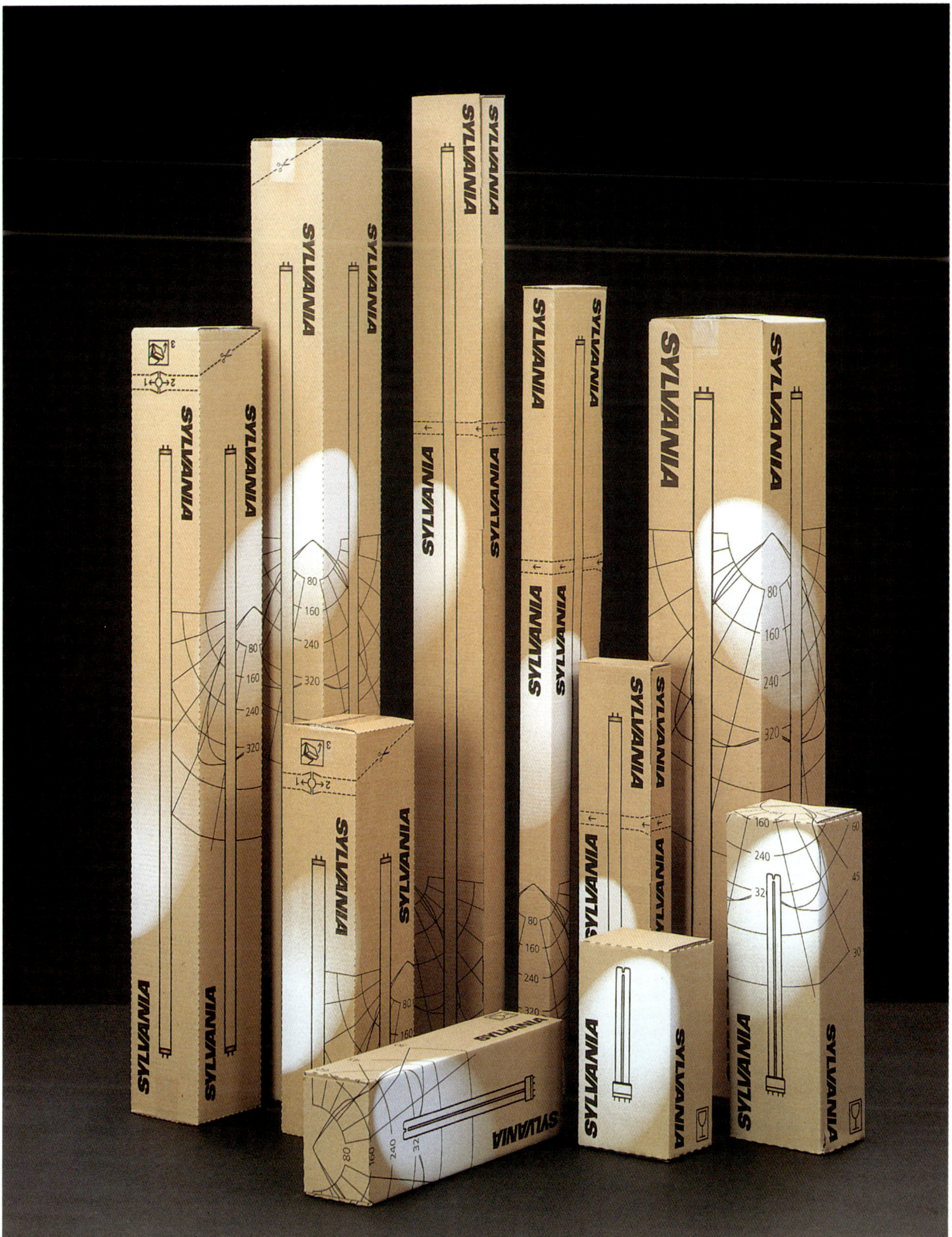
SYLVANIA
80
160
240
320
32
30
45
60

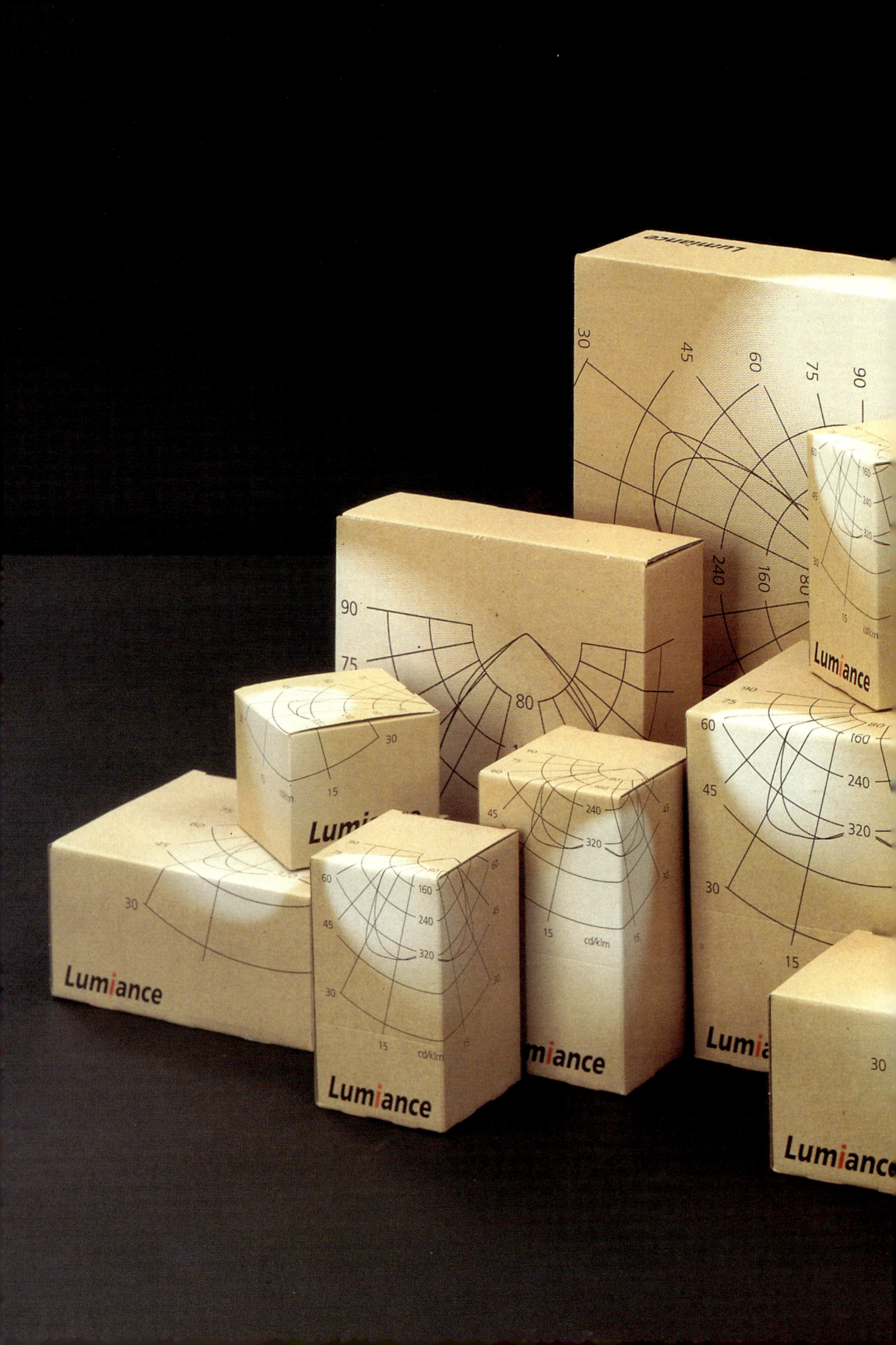
Lumiance
Lumiance
Lumiance
Lumiance

Lumiance

In commissioning a new design for wholesale products, Gert Boven, marketing manager for Lumiance, sought a motif that conveyed the company's commitment to providing total solutions in lighting. Premsela and Volk in Amsterdam came up with an abstract version of the polar curve diagram, used as a technical measursment of light output. To this they added a beam of white light. This design could be printed economically in two colors onto white recycled cardboard to create the packaging. The same design was used for all packaging: information on the individual lamps in each package was contained in a small white label fixed to each package. (In this way the same packaging could be used across the whole market.) The final design was unfussy in looks, economical to produce, and visually different from the packaging used by Sylvania's main competitors.

The plain wholesale packaging still used the light beam and polar curve consistently.

Elements

The new design encapsulates very effectively the company's approach to lighting (not just selling lamps and fittings but providing a complete service, including technical advice, planning and the design of special fixtures). As a result it is now being carried over as a corporate symbol into catalogs, manuals and promotional material. Creating a company logo from the packaging design upwards is an unusual approach, but it is practical in the context of a company like Sylvania, which needs to retain the customer loyalties of its different elements while moulding the separate units into a whole.

Our **packaging design** forms an integral part of our overall **corporate strategy.**

Lum*i*ance

The retail packaging (center) though much more direct about the product message, also uses the polar curve as a background image.

Designer	Premsela and Vonk, Amsterdam, The Netherlands
Cilent	Sylvania Lighting International, Geneva, Switzerland
Product	Electric Lamps
Material	Printed recycled card

Design Development

The packaging design's extension into catalogs can be seen below. The design for retail sales can be seen in the center. Here a different set of considerations apply. The customer is not so well informed, so an image of the lamp within appears on the outside of the pack. And to compete with other brands on the shelf, strong colors and typography are necessary. The polar curve, however, appears as a background image in this advertising photograph, to build a wider association between the individual product line and the company's overall strategy.

The beam and polar curve is extended from the main packaging to catalogues and reference material for professional use.

Olympic Weathering Stain

TECHNOLOGY

DESIGNER John Brady Design Consultants, Pittsburgh, PA, USA

CLIENT Olympic Paints and Stains, Pittsburgh, PA, USA

3   136/137

TYPE	Redesign of existing pack
PRODUCT	Exterior wood stain
MATERIAL	Printed paper label
SIZE	5 gallon tin
TIME	2 months
ELEMENTS	Barcode, user information and company logo
MARKET	Coastal property owners in NW USA and Canada
SUPPORT	Leaflets
KEY WORDS	Discriminating, aspirational, quality protection

Design Brief

The original design used a plain wooden background and a strongly informative, industrial look.

Weathering stain is a niche product. It is used mainly on natural wood house exteriors in coastal regions of the north-western USA and Canada. The paint is absorbed into the wood surface, protecting it from wind and salt – a process that takes six months or more, and produces a distinctive gray, bleached look to the woodwork. Within its market Olympic Weathering Stain was successful, but the client wanted to reposition it as the leading brand within the industry.

DESIGN The design problems posed by this brief stemmed from the nature of the product and its market. Because of the limited market, advertising was not a viable option to support the product: the label had to take the key role. The product was specialized, so it needed to carry considerable information for new consumers (and in parts of Canada this needed to be both in French and English). And in the design schedule, photography was not an option, as the product took six months to achieve the finished effect.

'*The brief looked for a design that maintained the context of the brand while bringing it to a wider market.*'

Key Factors

DESIGN
A single product can break ranks from a brand if it has a special market.
P.136

ELEMENTS
Special illustration can be used to visually explain a product to a new market.
P.139

SUPPORT
Continuity can be achieved by the use of logos and typefaces.
P.139

Protects Wood Beautif
OLY
Weathering Stain
Accelerates the natural weathering process and provides
a uniform, natural driftwood gray appearance
ONE GALLON
3.78 LITERS
DANGER! HARMFUL OR FATAL IF SWALLOWED. SEVERE SKIN AND EYE IRRITANT.
VAPOR HARMFUL. IRRITATES EYES, NOSE AND THROAT. COMBUSTIBLE.
See Back Panel for Precautions and First Aid.

TECHNOLOGY

Olympic – design development

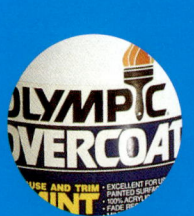

3 | 138/139

The brief looked for a design that maintained the
context of the brand while bringing
it to a **wider market.**

Within the brand range, the Stain is distinguished by its illustrated label and linked by the logo.

ELEMENTS

Competing products sold themselves with an industrial look, stressing the practicality of the product: the previous design had used a similar theme, with a text-only label. Olympic decided to stress the product's contribution to natural weathering, and produce a more elegant design that would stand out on the shelf from competing products. So an illustration was commissioned from Martha Ann Booth, an artist based in Montara, California, whose work combined realism and romance. Since the paint is intended to be used on coastal properties, the illustration shows a beach house; to maximize the impact of the wooden structure, in the restricted space, beach and shoreline are reflected in the doors and windows.

The illustration brief was to show the paint after weathering (which takes several years), and to show the intended location; the sea and coast are reflected in the windows. A view across the sea would have been too distant to show the detailed effect, so this is a very neat compromise.

SUPPORT

The Olympic logo with its flaming paintbrush was retained as part of the new design, but otherwise the design was allowed to develop an independent look from the other products in the range since it was a niche product.

TECHNOLOGY

Keggy

DESIGNER In-house team
CLIENT Keggy Drink Systems GmbH, Neunkirchen, Germany

4   140/141

ELEMENTS	Identity number and client brewery or Keggy logos
PRODUCT	Pressure beer dispenser
MATERIAL	Stainless steel
SIZE	12.5 litre capacity
TIME	2/3 years
TYPE	New product
MARKET	General adult beer market
SUPPORT	Leaflets and mailings
KEY WORDS	Accessible, portable, fresh

Design Brief

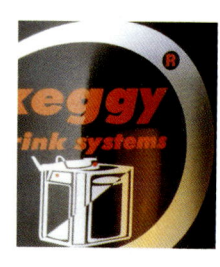

A new packaging design can arise from engineering innovation, as well as from design ideas. Keggy is a good example of this. Schäfer produce a range of pressurized liquid containers for industry, including larger barrels for brewers. Because the beer market in Germany is strongly disposed towards draught beer, they and their partner companies, Micro Matic and Rotarex, saw an opportunity for developing a product that could offer beer in smaller quantities than for a bar, to be used at a party at home or by a small club meeting. It could also be used for soft drinks and wine.

'Fresh draught drinks, every time and everywhere.'

Key Factors

DESIGN

A new product for a crossover market needs both a distinctive shape and name to make inroads.

P.142

ELEMENTS

Even a technological innovation needs to show it has advantages in subsidiary aspects of the placement.

P.142

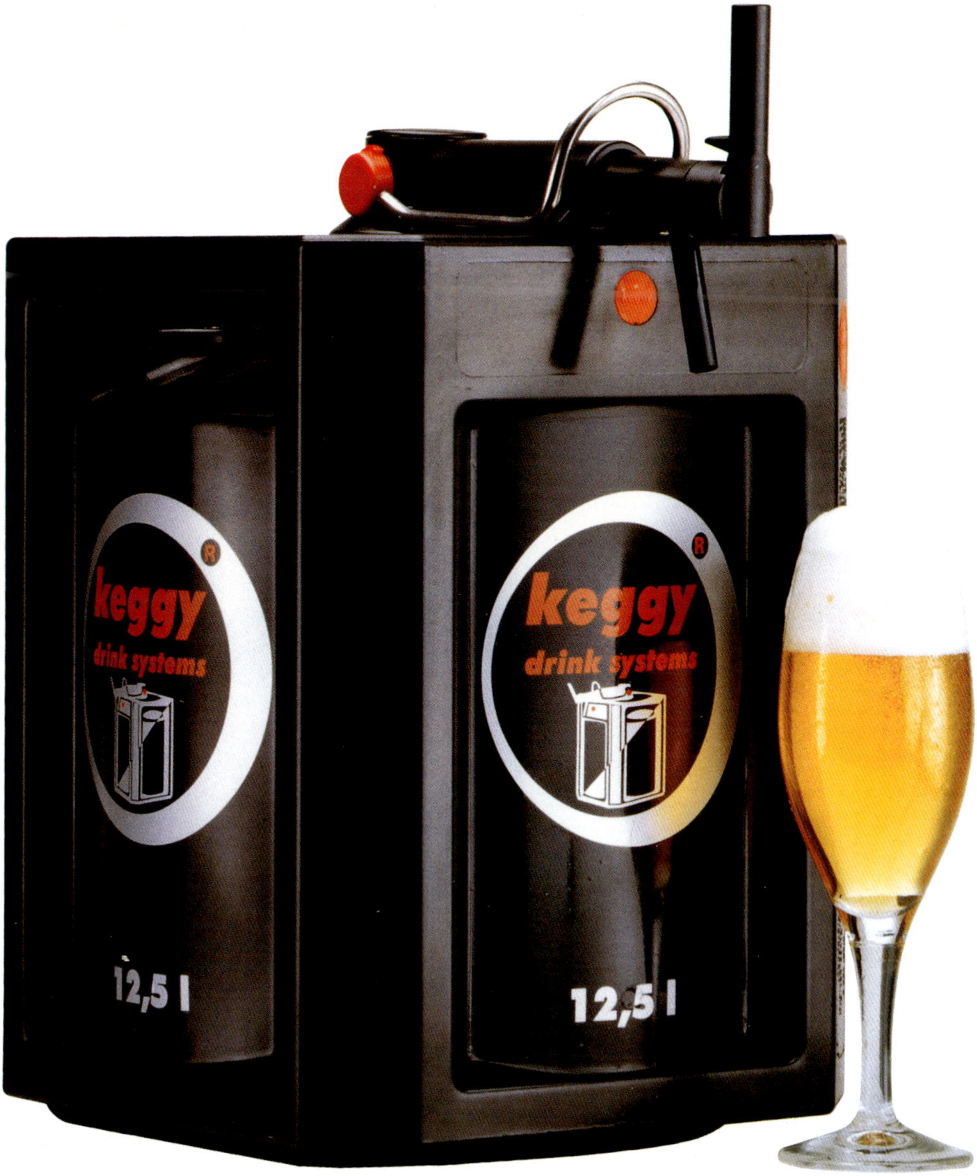
keggy
drink systems
12,5 l
keggy
drink systems
12,5 l

The engineer Hans Reichmann and his technical colleagues in the three firms working on Keggy set about developing this product. It needed to be easy to operate, without the maintenance and cleaning routines associated with complete keg systems. It had to keep beer fresh, even once first tapped, and deliver at the correct pressure all through the fill cycle. It had to be reusable, portable and easily stored. It had to be adaptable to carry the logos and colors of specific brewers, and meet safety requirements in its product range.

A special numbering system allows for stock control, especially for intermediate users offering a hire service.

ELEMENTS

The Keggy is an example of packaging innovation being led by engineering rather than by design. This has been the case in a number of important innovations, such as the Tetra pack, which keeps liquids fresh and free of contamination by the air vacuum. (Tetra maintain this commanding position by allowing the packs to be used only for goods for human consumption, so that consumers will not find a package on which they rely for milk or fruit juice appearing on shelves of bleach or shampoo.) Other examples are the chilled food packages which can be reheated safely in a microwave oven, and the anti-tamper devices now frequent on packaged foods. Consumers may not be aware of the innovations involved, but the successful packaging designer needs to keep up with the technology supporting the profession.

For professional users, the fact that the Keggy stores more efficiently is important.

Fresh draught
drinks everytime and
everywhere.

MATERIALS

The container itself is made from stainless steel with a polyurethane jacket. It has an integral carbon dioxide dispenser, and a simple, removable dispenser head in plastic. Additional features include cooling packs, a special single unit cooler, and a dedicated multiple pack refrigerator. The maximum use of internal space makes a compact yet practical product, which can be handled stacked on pallets or individually. As the products are reusable, each Keggy is coded with a permanent, indelible, machine-readable number which is indelibly foamed into the jacket. Logos can similarly be moulded into the sides of the Keggy, using a colored foil if required.

The logo itself can be used, or the barrel can be overprinted with a specific brand name.

The technical innovations need to be explained to assure consumers the product will perform as expected

Conclusion

The way packaging design has been described in the preceding pages has had a single focus. Not on the graphic values of design (though there are some great graphics), nor on structural packaging technicalities (though there again, there have been some important examples), nor on innovation (though I hope you've found novelty and excitement in the examples). The whole focus has been on the client/designer relationship, and the results that has produced. Every piece of packaging – good or bad, inappropriate or right-on, dull or brilliant – starts with a designer and a client meeting, and how well the design succeeds depends on the success of the relationship that starts at that meeting.

This is true of all design projects, of course, but it is particularly important in packaging where a whole range of considerations, outside the simple graphic remit of design, come into play. What is happening today in packaging design shows how designers are moving from the fringes of corporate activity towards the center (something that is also happening in corporate identity design, for example). There was a time when the client would ask for a design, the designers would suggest a couple of alternatives, and that was all. When Raymond Loewy redesigned the Lucky Strike cigarette pack in the 1930s, he had (according to his own account) one chatty meeting with the CEO, talking mainly about where they bought their shirts, and agreed a fee which was as much a bet as anything else (a bet Loewy won, of course...). Today a successful design must rest on an excellent understanding between designer and client, on a partnership between them, and on the designer understanding how the client corporation thinks and acts. This goes beyond a formal brief for a specific target, into a wider, and very important, new definition of design as a total communication process.

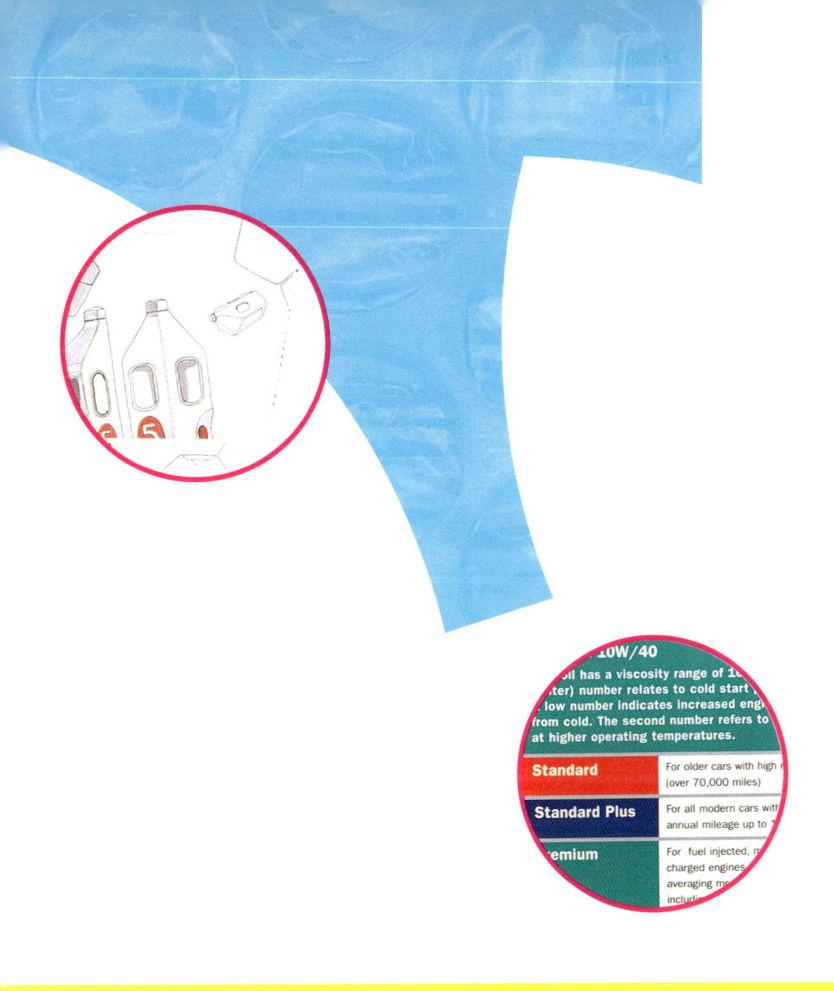

The final case study has therefore been chosen not for its design excellence (though it is excellent design), but for the way it illustrates how one of the best contemporary design companies approaches a new project, both in terms of design philosophy and wide client and market relations.

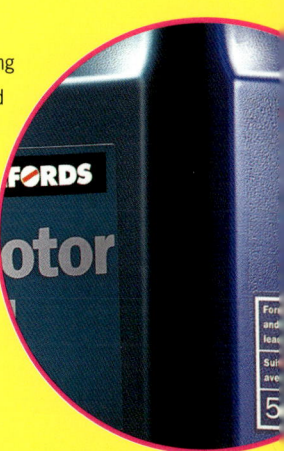

CONCLUSION

Halfords Motor Oil

DESIGNER Pentagram Design, and Lippa Pearce, London, UK
CLIENT Halfords plc, London, UK

5 146/147

The new design incorporates a specially placed handle and a visual oil gauge.

'The new design redefines and underlines the importance of the product to the client.'

TYPE	Redesign and rebranding of existing product
PRODUCT	Motor oil
MATERIAL	Blow moulded polythene
SIZE	5 litres and 1 litre
TIME	14 months (2 months design, 12 months development)
ELEMENTS	Logo, barcode, oil level indicator, descriptive label
MARKET	Retail through Halfords own stores
SUPPORT	In-store displays, leaflets
KEY WORDS	Ergonomic, efficient, reliable

Design Brief

Halfords are major retailers of motor accessories and cycles through their chain of UK shops. Their own-brand motor oil was an important product in sales terms, but market research suggested that it was not valued highly by customers and was not always used appropriately (many users were buying standard oil for cars requiring a superior grade, for example).

Halfords approached the international design group Pentagram with a brief to redesign the container, relabel it and reposition it in the market. The design practice Lippa Pearce were to handle the labelling. It was the first time Pentagram had worked for Halfords, though Lippa Pearce were already working with them on graphic design.

Key Factors

DEVELOPMENT
Paper sketches and bulk models, plus research and analysis, help the design process forward. 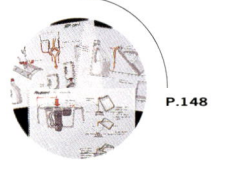
P.148

ELEMENTS
Clear labelling helps to communicate product specificity to the market. 
P.152

TYPE
A new ergonomic solution becomes a design equity. 
P.153

HALFORDS

Motor
Oil

Premium
10W/40

HALFORDS

Formulated to the same protection and performance standards as the leading brands

Suitable for all modern cars with average annual mileage

5 litre e

Halfords Motor Oil – design development

CONCLUSION

Rapid sketches make the visual analysis of a problem easier.

Research

The first stage was to make a series of drawings of existing and possible options. Such drawings serve to let the designer feel a way into the form. Thus Foskett and Thomson could explore the relationship between the three main elements; the spout, the main container and the handle. These had to function together in two main contexts: in carrying the can, and in pouring from it.

Bottle shapes, spout types and handle placements were the main focus of this phase.

The relationship between the handle and the main container was identified as a key issue, and so visually developed.

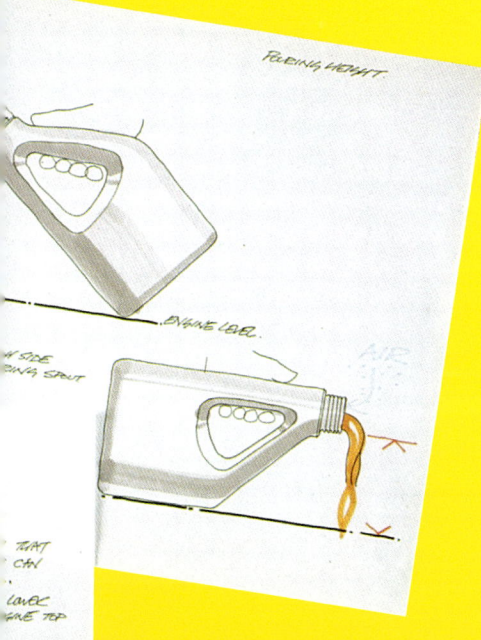

These drawings could be quite detailed, especially when looking at liquid movement.

The new design underlines the importance of the product to the client.

CONCLUSION

Halfords Motor Oil – design development

5 150/151

DESIGNER	Pentagram Design and Lippa Pearce, London, UK
CILENT	Halfords plc, London, UK
PRODUCT	Motor Oil
MATERIAL	Blow moulded polythene

DESIGN DEVELOPMENT

Peter Foskett and Gavin Thomson at Pentagram's product design section were asked to handle the brief. They began by analyzing the shape of oil cans and bottles, and soon realized that many of them were not easy to use, especially when pouring from a full five litre can into a narrow engine opening. A series of ergonomic studies suggested alternative handle positions would make this task easier and also create a distinctive shape. Frame and foam models of the three best candidates were made and analyzed in depth in comparison with two other standard forms and eight undeveloped prototypes and the final three were presented to the client.

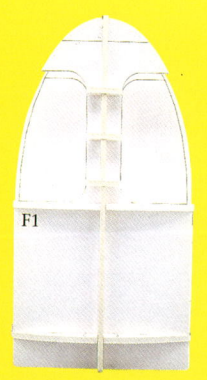

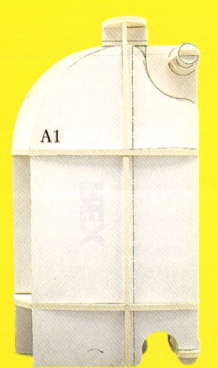

Three outline structural models were developed to check the capacity before bulk models (facing) were made.

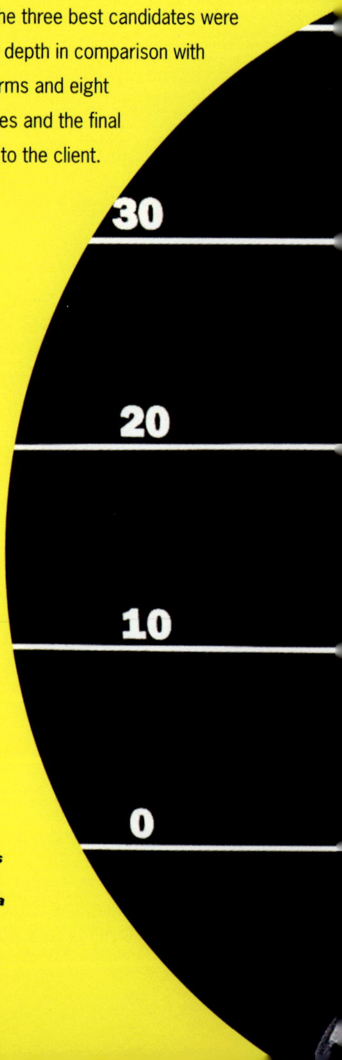

The angle of pour, as the can became empty, was seen as a critical factor.

30
20
10
0

Even a home-made model can serve to develop design understanding.

MATERIALS

Halfords chose the version with the handle in line with the spout. This was not the overall winner of the analysis, but it was visually original, especially in that consumer tests suggested its way of working was visually evident. (As soon as one picked up the bottle, it was clear that the hand positions controlled the flow of oil much more easily than with a conventional box can with off-centre spout.) But the design required pushing the manufacturing technology to its limits, as the design, incorporating a clear viewer for the oil level, meant expanding the material through a blow mould into a complex curved shape. The makers, Plysu, needed several months of trials to achieve a perfect result from the demanding specifications.

The label uses clear sans lettering with the Halfords main logo.

ELEMENTS

At the same time the question of inappropriate oil use needed to be addressed. To help customers understand clearly which oil to buy, a range of solid colors for the four grades and types of oil were fixed. The label design needed to explain the different grades clearly and directly. Lippa Pearce chose a sans serif face and simple descriptions to achieve this. These avoided slogans and jargon in favor of a direct explanation of the right use for the product.

Color coding and clear suitability guides are a key feature of the subsidiary labels.

Clear instructions on using the spout and can are part of the customer-friendly approach.

TYPE

The design process took two months, the development up to delivered products a further year. During that year, as Peter and Gavin point out, they were working alongside the client with outside manufacturers and suppliers to bring the product to market. 'It was like moving from the table across from the clients onto their side', they explain, 'while also learning about the whole process of oil production, and about Halfords' overall strategies and expectations.'

It was like moving from the table across from the clients onto their side.

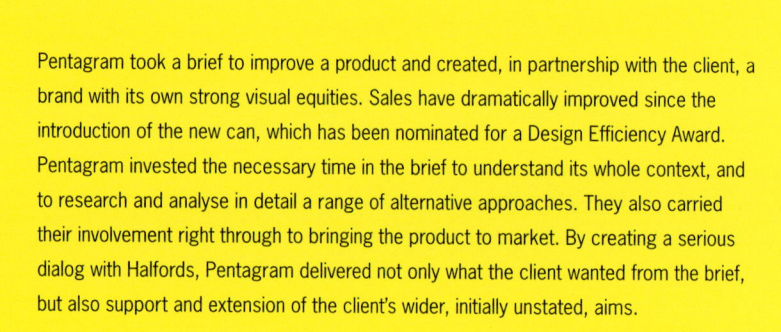

CONCLUSION

Halfords Motor Oil – design development

5   154/155

The clear graphics approach works equally well on the back and front of the design.

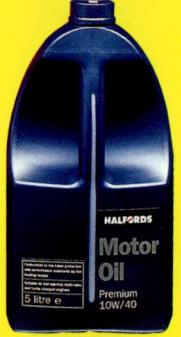

Pentagram took a brief to improve a product and created, in partnership with the client, a brand with its own strong visual equities. Sales have dramatically improved since the introduction of the new can, which has been nominated for a Design Efficiency Award. Pentagram invested the necessary time in the brief to understand its whole context, and to research and analyse in detail a range of alternative approaches. They also carried their involvement right through to bringing the product to market. By creating a serious dialog with Halfords, Pentagram delivered not only what the client wanted from the brief, but also support and extension of the client's wider, initially unstated, aims.

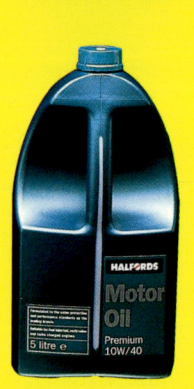

Conclusion

Pentagram's success was threefold: firstly, they met the immediate requirements of the brief; secondly, their in-depth study of the technical aspects of the brief created a new genre in the product range, which had physical equities that are of value to the client (the design is under patent processing at the time this is written); thirdly, they used their understanding of the client's overall business to reposition the product as a leading definition of Halfords' approach to the market. This was achieved through co-operation between designer and client. As Peter Foskett says, 'You have to be able to ask a client "why?"'.

The color range not only informs the user of oil categories, but has a strong presence in the shop.

Reading List

Blackwell, Lewis & Burney, Jan **The Retail Future**, Trefoil, London, 1992
A bright introduction to issues facing the retail market, with commentaries from Conran, Armani and others.

Caplan, Ralph **By Design**, Harper & Row, New York, 1981
The distilled design wisdom of one of America's finest design writers.

Glaser, Milton **Graphic Design**, New York, 1982
One of the best introductions to graphics, from a master designer.

Loewy, Raymond **Industrial Design**, New York, 1982
Raymond Loewy's own testament to his memorable career in design.

Lorenz, Christopher **The Design Dimension**, Blackwells, Oxford, 1993
An excellent explanation of the design issues facing business.

Milton, Howard **Packaging Design**, Design Council, London, 1991
A fine British analysis of packaging design issues.

Olins, Wally **Corporate Identity**, Thames & Hudson, London, 1995
A good analysis of current corporate identity practice.

Opie, R. **The Art of the Label**, Quarto, London, 1987
An anthology of historic and contemporary labels.

Pilditch, J **Winning Ways**, Harper & Row, London 1987
More on the design/business relationship, with case studies.

Wildbur, Peter **Information Graphics**, Trefoil, London 1989
A clear account of the challenges of information graphics.

Design Companies

Blueberry Design,
Unit 10/11, Design Center, Chelsea Harbour, London SW10, +44 171 351 3313

John Brady Design Consultancy Inc.,
17th Floor, Three Gateway Center, Pittsburgh, PA 15222-1012, (1) 412 288 9300

Alan Chan Design Co,
2/F Shui Lam Building, 23 Luard Rd, Wanchai, Hong Kong, (852) 2 2527 8228

Coy Design,
9520 Jefferson Blvd., Culver City, CA 90230, USA, +1 310 837 0173

Design Partners Incorporated,
338 Main St., Racine WI 534403, USA, +1 414 637 2233

De Witt Anthony Inc,
126 Main St., Northampton, Ma 01060, USA, +1 413 586 4304

DIN Design Co,
32 St Oswalds Place, London SE11 +44 171 582 0777

Graham Scott,
21 Circus Lane, Edinburgh, EH3 6SU, 0131 226 1550

Identica Partnership,
30 Queensway, London W2, +44 171 221 9900

Kan Creative Consultants,
Raapstraat 14, 2000 Antwerpen, Belgium, +32 (03)226 66 53

Lewis Moberly,
33 Gresse St., London W1P 2LP, +44 171 580 9252

Lippa Pearce Design,
358a Richmond Road, Twickenham TW1 2DU, (44) 181 744 2100

Raymond Loewy International,
30, Plympton St., London NW8 8AB, +44 171 402 8601

Pecos Design,
1106-A Payne Avenue, Austin, Texas 78757, (1) 512 302 5719

Pentagram Design Ltd.,
11 Needham Rd, London W11, +44 171 229 3477

BRS Premsela Vonk,
Nieuwe Herengracht 89. 1018 VR Amsterdam, (31) 20 626 20 30

Kojitani, Irie & Inc.,
2-29-13 Minami Aoyama, Minato-ku, Tokyo 107, Japan, (0) 3 5486 4410

Wagstaffs Design Consultants,
147a Grosvenor Rd., London SW1, UK, +44 171 834 0534

Index

Adelphi Whisky 52–55
Alan Chan Design 86–87
Anthony De Witt 88–93
Benetton Tribu 96–101
Blueberry Design 28–29, 108–109
Bodyshop International 102–107
John Brady Design Consultants 136–139
Coy Design 62–65
Design Partners 66–71
DIN Design 74–75
Graham Scott 52–55
Guiltless Gourmet 38–41
Halfords Motor Oil 146–155
Horai Dairy Products 34–35
Identica Partnership 82–85
Insignia 108–109
Kan Design Consultants 118–121
Keggy Drink Systems GMBH 140–143
Koitani Irie & Inc 34–35
Leinenkugel's Brewery 66–69
Lewis Moberly 110–113
Lippa Pearce 155–159
Raymond Loewy International 30–33, 36–37, 56–59, 122–125
Lucozade 46–51
Metroblade 88–93
Nicole Fahri 74–75
Nicosia Creative Expresso 96–101
Olympic Weathering Stain 136–139
Orient Tea Rooms 86–87
Oropharma 118–121
Painter 128–129
Panem 36–37
Pecos Design 40–41
Pentagram Design 146–155
Premsela Vonk 130–135
Quady's Winery 62–65
Rucker Design Group 128–129
Safeway Premium Ice-cream 24–25
Safeway 16–25
Safeway Savers 18–21
Safeway Cyclon 22–23
Sainsbury's 44–45

Sellotape 82–85
Sempre 110–113
Simoniz 122–125
Smirnoff Vodka 56–59
Sylvania 130–135
Toblerone 28–29
Toilet Duck 116–117
Tomy 76–81
Virgin Cola 60–61
Wagstaffs 18–25
Yoplait 30–33

General Index

Brands 16–25, 34–35, 62–71, 76–85, 102–107,
   118–121, 130–135
Company Logos 16–25, 76–85, 102–107, 118–125,
   130–135
Links to advertising 50–51, 108–109
      personalities 60–61
      Websites 70–71
Materials:
      Cellophane 36–37
      Foil 38–41
      Glass 46–49, 52–59, 61–71, 110–113
      Gualapak 50–51
      Paper and card 28–29, 40–41, 56–59,
         66–71, 74–93, 96–101, 108–109,
         110–113, 130–135
Plastic 16–25, 30–33, 44–49, 60–61,
   96–107, 116–125, 146–155
Recycled/recyclable 96–101, 130–135
Metal 140–143
Tin plate 16–21, 128–129, 136–139
Tetrapack 16–21
Own-brand Products 16–25, 42–45, 102–107,
   110–113, 146–151
Stand-alone Products 36–37, 56–59, 108–109,
   116–117, 128–129